27.

D0205865

SOLAR CELLS
Operating Principles, Technology, and System Applications

PRENTICE-HALL SERIES IN SOLID STATE
PHYSICAL ELECTRONICS

Nick Holonyak, Jr., *Editor*

SOLAR CELLS
Operating Principles, Technology, and System Applications

MARTIN A. GREEN
University of New South Wales
Australia

Prentice-Hall, Inc., Englewood Cliffs, N.J. 07632

Library of Congress Cataloging in Publication Data

Green, Martin A.
  Solar cells.

  (Prentice-Hall series in solid state physical
electronics)
    Bibliography: p.
    Includes index.
    1. Solar cells. 2. Photovoltaic power generation.
I. Title. II. Series.
Tk2960.G73   621.31′244   81-4355
ISBN   0-13-822270-3   AACR2

To Judy and Brie

.312
57

Editorial/production supervision and interior design:
  BARBARA BERNSTEIN
Manufacturing buyer: JOYCE LEVATINO

Printed in the United States of America

10 9 8 7 6 5 4 3 2 1

PRENTICE-HALL INTERNATIONAL, INC., *London*
PRENTICE-HALL OF AUSTRALIA PTY. LIMITED, *Sydney*
PRENTICE-HALL OF CANADA, LTD., *Toronto*
PRENTICE-HALL OF INDIA PRIVATE LIMITED, *New Delhi*
PRENTICE-HALL OF JAPAN, INC., *Tokyo*
PRENTICE-HALL OF SOUTHEAST ASIA PTE. LTD., *Singapore*
WHITEHALL BOOKS LIMITED, *Wellington, New Zealand*

CONTENTS

v

This solar cell is made from a thin wafer of the semiconductor silicon, about 10 cm square and only a fraction of a millimeter thick. When the cell is illuminated, it converts the energy of the photons in the incident light into electrical energy. Under bright sunshine, the cell can supply a current of up to 3 A at a voltage of about $\frac{1}{2}$ V to an electrical load connected between the metallic contact grid apparent here and a second contact at the rear of the cell. (Photograph courtesy of Motorola, Inc.)

PREFACE

When sunlight strikes a solar cell, the incident energy is converted directly into electricity without any mechanical movement or polluting by-products. Far from being a laboratory curiosity, solar cells have been used for over two decades, initially for providing electrical power for spacecraft and more recently for terrestrial systems. There are very real prospects that the manufacturing technology for these cells can be improved dramatically in the near future. This would allow solar cells to be produced at prices where they could make significant contributions to world energy demands.

This book concentrates on providing descriptions of the basic operating principles and design of solar cells, of the technology used currently to produce cells and the improved technology soon to be in operation, and of considerations of importance in the design of systems utilizing these cells. Accordingly, the early chapters of the book review the properties of sunlight, the relevant properties of the semiconductor material from which the cells are constructed, and the interaction between these two elements. The next group of chapters treat in some detail the factors important in the design of solar cells, current technology for fabricating them, and probable technological developments in the future. The final chapters deal with system applications, ranging from the small systems commercially available at present to residential and central power systems that may be available in the future.

The book is intended primarily for the increasing numbers of engineers and scientists attracted to this rapidly expanding field. As such, it is suitable for use as a textbook for both undergraduate and graduate courses. A deliberate attempt has been made not to exclude the material contained within from those readers who are entering the field through a different route. For example, a rather pictorial review of the properties of semiconductors relevant to the understanding of solar cell operation is included. Although this may serve as a quick review for many readers, for other readers it may provide a framework on which the material in subsequent chapters can be supported. Irrespective of background, working through the text and associated exercises would place the reader in a very strong position for future activity in this area.

I would like to acknowledge the large number of people, too numerous to mention individually, who have stimulated my interest in solar cells over the last decade. I would particularly like to thank Andy Blakers, Bruce Godfrey, Phill Hart, and Mike Willison for their suggestions and indirect encouragement in this venture. Special thanks are due to Gelly Galang for her help in preparing the manuscript and to John Todd and Mike Willison for preparing photographs incorporated into the text. Finally, I would like to thank Judy Green for her support and encouragement during the fairly intense period in which this book was developed.

Martin A. Green

Chapter

1

SOLAR CELLS
AND SUNLIGHT

1.1 INTRODUCTION

Solar cells operate by converting sunlight directly into electricity using the electronic properties of a class of material known as semiconductors. In the following chapters, this elegant energy-conversion process will be examined starting from the basic physical principles of solar cell operation. From this basis, the mathematical equations quantifying the energy transformation are developed. This is followed by a description of the technology used to produce present commercial solar cells, based predominantly on a particular semiconductor, silicon. Improvements in this technology, as well as alternative technologies that hold the promise of significantly lower cost, are then described. Finally, the design of solar cell systems is discussed, ranging from small power supplies for remote-area use to possible future residential and central power-generating plants.

In this chapter, the history of solar cell development is outlined briefly, followed by a review of the properties of the sun and its radiation.

1.2 OUTLINE OF SOLAR CELL DEVELOPMENT

Solar cells depend upon the *photovoltaic effect* for their operation. This effect was reported initially in 1839 by Becquerel, who observed a light-dependent voltage between electrodes immersed in an electrolyte. It was observed in an all-solid-state system in 1876 for the case of selenium. This was followed by the development of photocells based on both this material and cuprous oxide. Although a silicon cell was reported in 1941, it was not until 1954 that the forerunner of present silicon cells was announced. This device represented a major development because it was the first photovoltaic structure that converted light to electricity with reasonable efficiency. These cells found application as power sources in spacecraft as early as 1958. By the early 1960s, the design of cells for space use had stabilized, and over the next decade, this was their major application. Reference 1.1 is a good source of more detailed material up to this stage.

The early 1970s saw an innovative period in silicon cell development, with marked increases in realizable energy-conversion efficiencies. At about the same time, there was a reawakening of interest in terrestrial use of these devices. By the end of the 1970s, the volume of cells produced for terrestrial use had completely outstripped that for space use. This increase in production volume was accompanied by a significant reduction in solar cell costs. The early 1980s saw newer device technologies being evaluated at the pilot production stage, poised to enable further reduction in costs over the coming decade. With such cost reductions, a continual expansion of the range of commercial applications is ensured for this approach to utilizing the sun's energy.

1.3 PHYSICAL SOURCE OF SUNLIGHT

Radiant energy from the sun is vital for life on our planet. It determines the surface temperature of the earth as well as supplying virtually all the energy for natural processes both on its surface and in the atmosphere.

The sun is essentially a sphere of gas heated by a nuclear fusion reaction at its center. Hot bodies emit electromagnetic radiation with a wavelength or spectral distribution determined by the body's temperature. For a perfectly absorbing or "black" body, the spectral distribution of the emitted radiation is given by *Planck's radiation law* (Ref. 1.2). As indicated in Fig. 1.1, this law indicates that as a body is heated, not only does the total energy of the electromagnetic

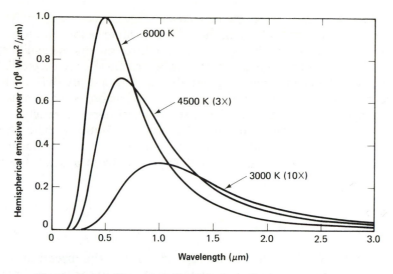

Figure 1.1. Planckian black-body radiation distributions for different black-body temperatures.

radiation emitted increase, but the wavelength of peak emission decreases. An example of this within most of our ranges of experience is that metal, when heated, glows red and then yellow as it gets hotter.

Temperatures near the sun's center are estimated to reach a warm 20,000,000 K. However, this is *not* the temperature that determines the characteristic electromagnetic radiation emission from the sun. Most of the intense radiation from the sun's deep interior is absorbed by a layer of negative hydrogen ions near the sun's surface.

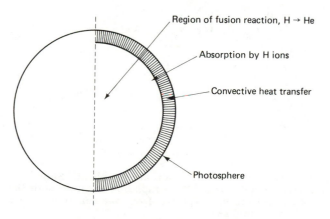

Figure 1.2. Principal features of the sun.

3

These ions act as continuous absorbers over a great range of wavelengths. The accumulation of heat in this layer sets up convective currents that transport the excess energy through the optical barrier (Fig. 1.2). Once through most of this layer, the energy is reradiated into the relatively transparent gases above. The sharply defined level where convective transport gives way to radiation is known as the *photosphere*. Temperatures within the photosphere are much cooler than at the sun's interior but are still a very high 6000 K. The photosphere radiates an essentially continuous spectrum of electromagnetic radiation closely approximating that expected from a black body at this temperature.

1.4 THE SOLAR CONSTANT

The radiant power per unit area perpendicular to the direction of the sun outside the earth's atmosphere but at the mean earth-sun distance is essentially constant. This radiation intensity is referred to as the *solar constant* or, alternatively, *air mass zero* (AM0) *radiation*, for reasons that will soon become apparent.

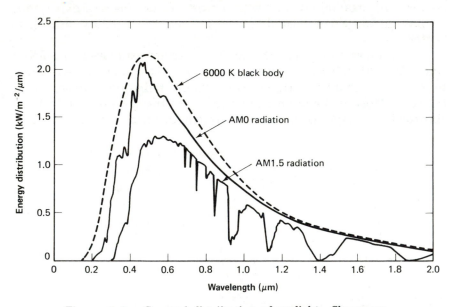

Figure 1.3. Spectral distribution of sunlight. Shown are the cases of AM0 and AM1.5 radiation together with the radiation distribution expected from the sun if it were a black body at 6000 K.

The presently accepted value of the solar constant in photovoltaic work is 1.353 kW/m^2. This value has been determined by taking a weighted average of measurements made by equipment mounted on balloons, high-altitude aircraft, and spacecraft (Ref. 1.3). As indicated by the two uppermost curves in Fig. 1.3, the spectral distribution of AM0 radiation differs from that of an ideal black body. This is due to such effects as differing transmissivity of the sun's atmosphere at different wavelengths. Currently accepted values for this distribution are tabulated in Ref. 1.3. A knowledge of the exact distribution of the energy content in sunlight is important in solar cell work because these cells respond differently to different wavelengths of light.

1.5 SOLAR INTENSITY AT THE EARTH'S SURFACE

Sunlight is attenuated by at least 30% during its passage through the earth's atmosphere. Causes of such attenuation are (Ref. 1.4):

1. Rayleigh scattering or scattering by molecules in the atmosphere. This mechanism attenuates sunlight at all wavelengths but is most effective at short wavelengths.
2. Scattering by aerosols and dust particles.
3. Absorption by the atmosphere and its constituent gases—oxygen, ozone, water vapor, and carbon dioxide, in particular.

A typical spectral distribution of sunlight reaching the earth's surface is shown by the lower curve of Fig. 1.3, which also indicates the absorption bands associated with molecular absorption.

The degree of attenuation is highly variable. The most important parameter determining the total incident power under clear conditions is the length of the light path through the atmosphere. This is shortest when the sun is directly overhead. The ratio of any actual path length to this minimum value is known as the *optical air mass*. When the sun is directly overhead, the optical air mass is unity and the radiation is described as *air mass one* (AM1) *radiation*. When the sun is an angle θ to overhead, the air mass is given by

$$\text{Air mass} = \frac{1}{\cos \theta} \qquad (1.1)$$

Hence, when the sun is 60° off overhead, the radiation is AM2. The easiest way to estimate the air mass is to measure the length of the shadow s cast by a vertical structure of height h. Then

$$\text{Air mass} = \sqrt{1 + \left(\frac{s}{h}\right)^2} \qquad (1.2)$$

With increasing air mass but with other atmospheric variables constant, the energy reaching the earth is attenuated at all wavelengths, with attenuation in the vicinity of the absorption bands of Fig. 1.3 becoming even more severe.

Hence, as opposed to the situation outside the earth's atmosphere, terrestrial sunlight varies greatly both in intensity and spectral composition. To allow meaningful comparison between the performances of different solar cells tested at different locations, a terrestrial standard has to be defined and measurements referred to this standard. Although the situation is in a state of flux, the most widely used terrestrial standard at the time of writing is the AM1.5 distribution of Table 1.1, also plotted as the terrestrial curve in Fig. 1.3. In the photovoltaic program of the U.S. government, this distribution, essentially scaled up so that the total power density content is 1 kW/m^2, was incorporated as a standard in 1977 (Ref. 1.5). The latter power density is close to the maximum received at the earth's surface.

1.6 DIRECT AND DIFFUSE RADIATION

The composition of terrestrial sunlight is further complicated by the fact that, as well as the component of radiation directly from the sun, atmospheric scattering gives rise to a significant indirect or *diffuse* component. Even in clear, cloudless skies, the diffuse component can account for 10 to 20% of the total radiation received by a horizontal surface during the day.

For less sunny days, the percentage of radiation on a horizontal surface that is diffuse generally increases. From observed data (Ref. 1.6), the following statistical trends can be discerned. For days on which there is a notable lack of sunshine, most of the radiation will be diffuse. This will be true, in general, for days on which the total radiation received is up to one-third that which would be received on a clear, sunny day at the same time of the year. For days between the sunny and cloudy extremes mentioned above, where about one-half of clear-day radiation is received, about 50% of this generally will be diffuse. Poor weather will not only cause some

Table 1.1 Solar Spectrum—Air Mass 1.5*

Wavelength (μm)	W/(m² - μm)	Wavelength (μm)	W/(m² - μm)	Wavelength (μm)	W/(m² - μm)	Wavelength (μm)	W/(m² - μm)	Wavelength (μm)	W/(m² - μm)
0.295	0	0.595	1262.61	0.870	843.02	1.276	344.11	2.388	31.93
0.305	1.32	0.605	1261.79	0.875	835.10	1.288	345.69	2.415	28.10
0.315	20.96	0.615	1255.43	0.8875	817.12	1.314	284.24	2.453	24.96
0.325	113.48	0.625	1240.19	0.900	807.83	1.335	175.28	2.494	15.82
0.335	182.23	0.635	1243.79	0.9075	793.87	1.384	2.42	2.537	2.59
0.345	234.43	0.645	1233.96	0.915	778.97	1.432	30.06		
0.355	286.01	0.655	1188.32	0.925	217.12	1.457	67.14		
0.365	355.88	0.665	1228.40	0.930	163.72	1.472	59.89		
0.375	386.80	0.675	1210.08	0.940	249.12	1.542	240.85		
0.385	381.78	0.685	1200.72	0.950	231.30	1.572	226.14		
0.395	492.18	0.695	1181.24	0.955	255.61	1.599	220.46		
0.405	751.72	0.6983	973.53	0.965	279.69	1.608	211.76		
0.415	822.45	0.700	1173.31	0.975	529.64	1.626	211.26		
0.425	842.26	0.710	1152.70	0.985	496.64	1.644	201.85		
0.435	890.55	0.720	1133.83	1.018	585.03	1.650	199.68		
0.445	1077.07	0.7277	974.30	1.082	486.20	1.676	180.50		
0.455	1162.43	0.730	1110.93	1.094	448.74	1.732	161.59		
0.465	1180.61	0.740	1086.44	1.098	486.72	1.782	136.65		
0.475	1212.72	0.750	1070.44	1.101	500.57	1.862	2.01		
0.485	1180.43	0.7621	733.08	1.128	100.86	1.955	39.43		
0.495	1253.83	0.770	1036.01	1.131	116.87	2.008	72.58		
0.505	1242.28	0.780	1018.42	1.137	108.68	2.014	80.01		
0.515	1211.01	0.790	1003.58	1.144	155.44	2.057	72.57		
0.525	1244.87	0.800	988.11	1.147	139.19	2.124	70.29		
0.535	1299.51	0.8059	860.28	1.178	374.29	2.156	64.76		
0.545	1273.47	0.825	932.74	1.189	383.37	2.201	68.29		
0.555	1276.14	0.830	923.87	1.193	424.85	2.266	62.52		
0.565	1277.74	0.835	914.95	1.222	382.57	2.320	57.03		
0.575	1292.51	0.8465	407.11	1.236	383.81	2.338	53.57		
0.585	1284.55	0.860	857.46	1.264	323.88	2.356	50.01		

*Total energy content = 832 W/m².
Source: Ref. 1.5.

7

regions of the world to receive low levels of solar radiation but will also cause a significant proportion of it to be diffuse.

Diffuse sunlight generally has a different spectral composition from direct sunlight. Generally, it will be richer in the shorter or "blue" wavelengths, giving rise to further variability in the spectral composition of light received by a solar cell system. Uncertainty in the distribution of diffuse radiation from different directions in the sky introduces other uncertainties when calculating radiation levels on inclined surfaces from data generally recorded on horizontal surfaces. A common assumption is that diffuse light is isotropic (uniform in all directions), although the region of the sky surrounding the sun is the most intense source of this radiation.

Photovoltaic systems based on concentrated sunlight can generally only accept rays spanning a limited range of angles. Hence, they usually have to track the sun to utilize the direct component of sunlight, with the diffuse component wasted. This tends to offset the advantage gained by such tracking systems of intercepting maximum power density by always being normal to the sun's rays.

1.7 APPARENT MOTION OF THE SUN

The earth spins daily on an imaginary axis orientated in a fixed direction relative to the plane of the earth's yearly orbit about the sun. The angle this direction makes with the orbital plane is the solar declination ($23°27'$). Perhaps less familiar are the details of the apparent motion of the sun relative to a fixed observer on earth resulting from the relationship described above.

This apparent motion is indicated in Fig. 1.4 for an observer at latitude $35°$ north. On any given day, the plane of the sun's apparent orbit lies at an angle equal to the latitude from the observer's vertical. At the equinoxes (March 21 and September 23), the sun rises due east and sets due west, so that the altitude of the sun at solar noon on these days equals $90°$ minus the latitude. At the summer and winter solstices (June 21 and December 22, respectively, for the northern hemisphere, the opposite for the southern), the altitude at solar noon has increased or decreased by the declination of the earth ($23°27'$).

1.8 SOLAR INSOLATION DATA

The ideal situation for the design of photovoltaic systems would be when there were detailed records of the solar insolation at the site

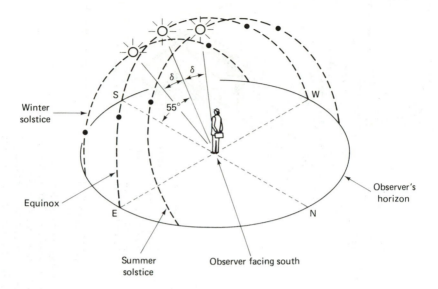

δ = declination of the earth = 23°27'

Figure 1.4. Apparent motion of the sun relative to a fixed observer at latitude 35° in the northern hemisphere. The path of the sun is shown at the equinoxes and the summer and winter solstices. The position of the sun is shown at solar noon on each of these days. The shaded circles represent the sun's position 3 h before and after solar noon.

selected for installation. Not only would data on the direct and diffuse components of light be desirable, but data on corresponding ambient temperatures as well as wind speed and direction could be used to advantage. Although there are stations at various locations around the world that do monitor all these parameters, present economies favor the use of photovoltaic systems in remote regions of the world where it is unlikely that such information is available.

The available insolation at a given location depends not only on gross geographical features such as latitude, altitude, climatic classification, and prevailing vegetation, but it also depends strongly upon local geographical features. Although unable to incorporate the latter features, maps of solar insolation distribution are available for different parts of the world. These have usually been prepared by combining measured insolation data with data estimated from a large network of stations around the world monitoring hours of sunshine.

The information most generally available is the average daily total or *global radiation* on a horizontal surface. A widely used source

9

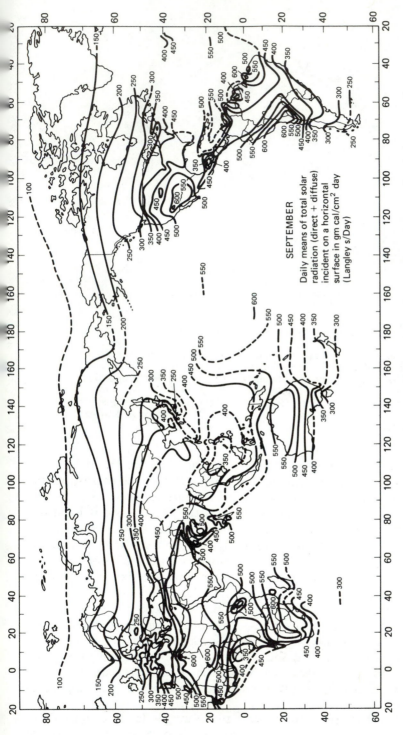

Figure 1.5. Worldwide distribution of solar energy during the month of September. The contours are daily means of global solar radiation incident on a horizontal surface expressed in *langleys*. A langley equals 1 cal/cm². To convert to MJ/m², multiply by 0.0418. To convert to kWh/m², multiply by 0.0116. The distribution of solar energy during September is roughly indicative of annual average daily radiation for a given site. Similar curves for other months of the year appear in Ref. 1.7. (After Ref. 1.7.)

for such data is Ref. 1.7. This lists, for each month of the year, average daily global radiation on a horizontal surface for hundreds of insolation monitoring stations around the world. It also lists this information estimated from sunshine-hour records, taking into account climatic and vegetation data for several hundred other locations. This information has been incorporated into a sequence of world maps showing contours of constant insolation for each month of the year. Such contours are illustrated in Fig. 1.5 for a month of equinox, September. This month corresponds approximately to average insolation levels throughout the year for most locations.

1.9 SUMMARY

Although sunlight outside the earth's atmosphere is relatively constant, the situation at the earth's surface is more complex. Terrestrial sunlight varies dramatically and unpredictably in availability, intensity, and spectral composition. On clear days, the length of the sunlight's path through the atmosphere or the optical air mass is an important parameter. The indirect or diffuse component of sunlight can be particularly important for less ideal conditions. Reasonable estimates of global radiation (direct plus diffuse) received annually on horizontal surfaces are available for most regions of the world. However, there are uncertainties involved in using this for a specific site because of the large deviations that can be caused by local geographical conditions and approximations involved in conversion to radiation on inclined surfaces.

EXERCISES

1.1. Estimate the solar constant for Mercury and Mars given that the mean distances from the sun to Earth, Mercury, and Mars are 150, 58, and 228 million kilometers, respectively.

1.2. The sun is at an altitude of $30°$ to the horizontal. What is the corresponding air mass?

1.3. Calculate the sun's altitude at solar noon on June 21 at Sydney (latitude $34°S$), San Francisco (latitude $38°N$), and New Delhi (latitude $29°N$).

1.4. The global radiation at solar noon on a summer solstice in Albuquerque, New Mexico (latitude $35°N$), is 60 mW/cm^2. Assume that 30% of this is diffuse radiation and make the approximations that the ground surrounding the module is nonreflecting and the diffuse radiation is uniformly distributed across the sky. Estimate the radiation intensity on a flat surface facing south at an angle of 45° to the horizontal.

REFERENCES

[1.1] M. WOLF, "Historical Development of Solar Cells," in *Solar Cells*, ed. C. E. Backus (New York: IEEE Press, 1976).

[1.2] R. SIEGEL AND J. R. HOWELL, *Thermal Radiation Heat Transfer* (New York: McGraw-Hill, 1972).

[1.3] M. P. THEKACKARA, *The Solar Constant and the Solar Spectrum Measured from a Research Aircraft*, NASA Technical Report No. R-351, 1970.

[1.4] P. R. GAST, "Solar Radiation," in *Handbook of Geophysics*, ed. C. F. Campen et al. (New York: Macmillan, 1960), pp. 14-16 to 16-30.

[1.5] *Terrestrial Photovoltaic Measurement Procedures*, Report ERDA/NASA/1022-77/16, June 1977.

[1.6] B. Y. LIU AND R. C. JORDAN, "The Interrelationship and Characteristic Distribution of Direct, Diffuse and Total Solar Radiation," *Solar Energy* 4 (July 1960), 1-19.

[1.7] G. O. G. LÖF, J. A. DUFFIE, AND C. O. SMITH, *World Distribution of Solar Energy*, Report No. 21, Solar Energy Laboratory, University of Wisconsin, July 1966.

Chapter

2

REVIEW
OF SEMICONDUCTOR
PROPERTIES

2.1 INTRODUCTION

In Chapter 1, the properties of sunlight were reviewed. It is now appropriate to look at the properties of the other important component in photovoltaic solar energy conversion, semiconducting material.

The aim of this chapter is *not* to treat the properties of semiconductors rigorously from fundamentals. Rather, it is to highlight those properties of semiconductors that are important in the design and operation of solar cells. As such, the chapter may suffice for quick revision for readers already acquainted with these properties while containing adequate information to allow those not as well acquainted to establish a framework on which subsequent material can be supported. To strengthen this framework, readers in the latter category are referred to one of the many textbooks specifically directed at treating semiconductor properties more fundamentally (Refs. 2.1 to 2.4).

2.2 CRYSTAL STRUCTURE AND ORIENTATIONS

Most of the photovoltaic materials described in this book are *crystalline*, at least on a microscopic scale. Ideal crystalline material is characterized by an orderly, periodic arrangement of the atoms of which it is composed.

In such an orderly arrangement, it is possible to build up the entire crystal structure by repeatedly stacking a small subsection. The smallest such section with which this is possible is known as a *primitive cell*. Such primitive cells naturally contain all the information required to reconstruct the locations of atoms in the crystal but often have awkward shapes. Consequently, it can be more convenient to work with a larger *unit cell* which also contains this information but generally has a simpler shape. For example, Fig.2.1(a) shows the unit cell for an atomic arrangement known as face-centered-cubic and Fig. 2.1(b) shows the corresponding primitive cell. The directions defining the outline of the unit cell are orthogonal, whereas this is not the case for the primitive cell. The length of the edge of the unit cell is known as the *lattice constant*.

The orientation of planes within the crystal can be expressed in terms of the unit cell structure by using the system of *Miller indices*. The vectors defining the outline of the unit cell are used as the basis of a coordinate system as in Fig. 2.1(a). A plane of the orientation in question is imagined passing through the origin of the coordinate system. Then the next plane parallel to this which passes through atomic sites along each of the coordinate axis is considered. An ex-

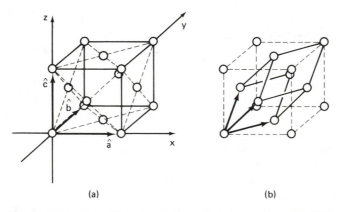

(a)   (b)

Figure 2.1.   (a)   Unit   cell   for   the   face-centered-cubic atomic arrangement.   The unit cell is selected in this case so that the directions defining its outline are orthogonal.   The vectors $\hat{a}$, $\hat{b}$, and $\hat{c}$ are unit vectors in each of these directions.   (b)   Primitive cell for the same atomic arrangement.

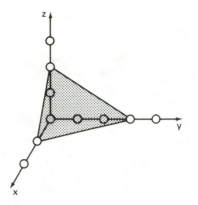

Figure 2.2. Sketch of a plane in a crystal described by the Miller indices (6 2 3).

ample is shown in Fig. 2.2. The intercepts in this case along each of the axes are 1, 3, and 2 atoms from the origin. Taking inverses gives 1, $\frac{1}{3}$, and $\frac{1}{2}$. The smallest integrals with the same ratio are 6, 2, and 3. This plane is then expressed in Miller indices as the (6 2 3) plane. Negative intercepts are indicated by a bar over the top of the corresponding index (e.g., – 2 would be written as $\bar{2}$).

Directions within the crystal are expressed in a condensed form of vector notation. A vector in the direction of interest is scaled so it is expressed in the form $h\,\hat{a} + k\,\hat{b} + l\hat{c}$, where $\hat{a}$, $\hat{b}$, and $\hat{c}$ are unit vectors along each of the axes of the coordinate system as in Fig. 2.1(a) and h, k, and l are integers. This direction is then described as the $[h\ k\ l]$ direction. The use of square brackets distinguishes directions from Miller indices. Note that for cubic unit cells, the $[h\ k\ l]$ direction is perpendicular to the $(h\ k\ l)$ plane.

Finally, there are planes within the crystal structure which are equivalent. For example, for the face-centered-cubic lattice of Fig. 2.1(a), differences between the (100), (010), and (001) planes depend only on the choice of origin. Collectively, the corresponding set of equivalent planes is known as the {100} set, with braces reserved for this use.

Figure 2.3(a) shows the atomic arrangement found in many of the semiconductors important in solar cell technology. This is the arrangement for silicon (Si) crystals as well as for crystals of gallium arsenide (GaAs) and cadmium sulfide (CdS). The latter are *compound* semiconductors involving more than one type of atom in the crystal structure. The arrangement shown is generally referred to as the *diamond lattice* or *zincblende lattice* (for compound semiconductors such as GaAs). The unit cell is cubic, as indicated. Figure

15

(a)

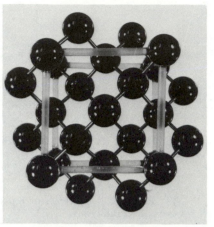
(b)

(c)

(d)

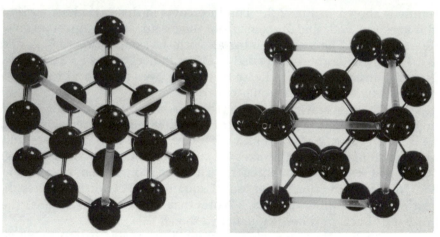

Figure 2.3. (a) Model of the diamond lattice which indicates the atomic structure of many of the semiconductors
of importance in solar cells. Also shown is the outline of
the unit cell. (b) View of the same structure looking in
the [100] direction. (c), (d) Views in the [111] and
[110] directions, respectively.

2.3(b)-(d) show atomic arrangements looking at the lattice from selected directions. These highlight the substantial differences in the
physical arrangement of atoms in different directions which give rise
to directional variations of importance in solar cell work (e.g., see
Exercise 2.2).

16

2.3 FORBIDDEN ENERGY GAPS

An electron in free space has an essentially continuous range of energy values that it can attain. The situation in a crystal can be quite different.

Electrons associated with isolated atoms have a well-defined set of discrete energy levels available to them. As several atoms are brought closer together, the original levels spread out into bands of allowed energy as indicated in Fig. 2.4. When the atoms are in ordered arrangements as in crystals, there will be characteristic distances between them. Figure 2.4(a) shows the case of a crystal where the characteristic separation between atoms, d, is such that the crystal has bands of energies allowed to electrons (corresponding to the atomic energy levels) separated by bands of forbidden energy. A different situation is shown in Fig. 2.4(b), where the bands have overlapped to give virtually a continuum of allowed energies at the value of d characteristic of a different crystalline material.

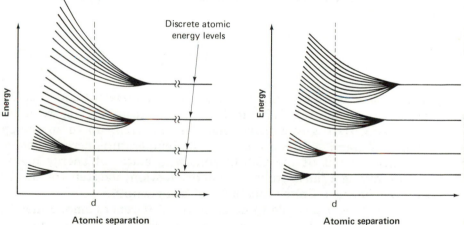

Figure 2.4. Schematic indicating how the discrete energies allowed to electrons in an isolated atom split up into bands of allowed energies when a number of similar atoms are brought together in a crystal:

(a) For this case, d, the characteristic spacing of atoms in a crystal, is such that there are bands of energies allowed to electrons separated by bands of forbidden energy.

(b) In this case, d is such that the uppermost bands have overlapped.

2.4 PROBABILITY OF OCCUPATION
OF ALLOWED STATES

At low temperatures, electrons in a crystal occupy the lowest possible energy states.

At first sight, it might be expected that the equilibrium state of a crystal would be one in which the electrons are all in the lowest allowed energy level. However, this is *not* the case. A fundamental physical theorem, the *Pauli exclusion principle*, implies that each allowed energy level can be occupied by, at most, two electrons, each of opposite "spin." This means that, at low temperatures, all available states in the crystal up to a certain energy level will be occupied by two electrons. This energy level is known as the *Fermi level* (E_F).

As the temperature increases, some electrons gain energy in excess of the Fermi level. The probability of occupation of an allowed electron state of any given energy E can be calculated from statistical considerations for this more general case, taking into account the constraints imposed by the Pauli exclusion principle (Refs. 2.1 to 2.4). The result is the *Fermi–Dirac distribution function* $f(E)$, given by

$$f(E) = \frac{1}{1 + e^{(E - E_F)/k\,T}} \tag{2.1}$$

where k is a constant known as *Boltzmann's constant* and T is the absolute temperature. This function is plotted in Fig. 2.5. Near absolute zero, $f(E)$ is essentially unity, as expected, up to an energy equal to E_F, and zero above E_F. As the temperature increases, there is a smearing out of the distribution, with states of energy higher than E_F having a finite probability of occupation, and states of energy below E_F having a finite probability of being empty.

It is now possible to describe the differences among metals, insulators, and semiconductors in terms of electronic band structure. Metals have an electronic structure such that E_F lies within an allowed band (Fig. 2.6). The cause of this may be that there are insufficient electrons available to fill an available band if the band structure is as shown in Fig. 2.4(a), or alternatively, that there are overlapping bands as in Fig. 2.4(b). Insulators have one band fully occupied by electrons and a large energy gap between this band and the next highest band, which is devoid of electrons at low temperatures. From the discussion in the earlier part of this section, it follows that E_F must lie within the forbidden band (Fig. 2.6).

A band in which there are no electrons obviously cannot make

18

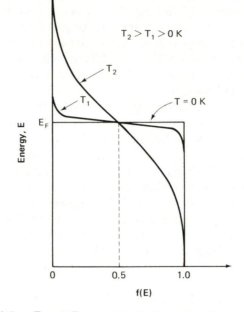

Figure 2.5.   Fermi–Dirac   distribution   function.   States
above   the   Fermi   level, E_F,   have   a   low   probability   of   being
occupied   by   electrons,   whereas   those   below   are   likely   to   be
so   occupied.

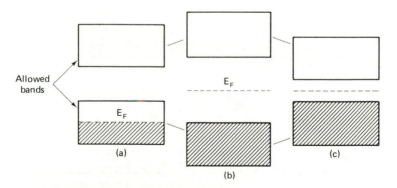

Figure 2.6.   Diagrams   showing   the   way   in   which   allowed
states are occupied by electrons in:

(a)   A metal.
(b)   An insulator.
(c)   A semiconductor.

19

any contribution to current flow in the crystal. More surprisingly, *neither can a completely full band.* To contribute to such flow, an electron must extract energy from the applied field. In a completely filled band, this is not possible. There are no vacant allowed energy levels in the vicinity into which an electron can be excited. Hence, an insulator does not conduct electricity, whereas a metal, with an abundance of such levels, does.

A semiconductor is just an insulator with a narrow forbidden band gap. At low temperatures, it does not conduct. At higher temperatures, there is sufficient smearing out of the Fermi–Dirac distribution function to ensure that some levels in the originally completely filled band (*valence band*) are now vacant and some in the next-highest band (*conduction band*) are occupied. The electrons in the conduction band, with an abundance of unoccupied energy states in the vicinity, can contribute to current flows. Since there are now unoccupied levels in the valence band, an additional contribution also comes from electrons in this band.

2.5 ELECTRONS AND HOLES

A very simple but reasonably good analogy to current flow processes in a semiconductor is provided by an idealized two-level parking station (Fig. 2.7). Consider the case where the bottom level of this station is completely filled with cars and the top level completely empty, as in Fig. 2.7(a). Then there is no room for any car to move. If one car is moved from the first to the second level as

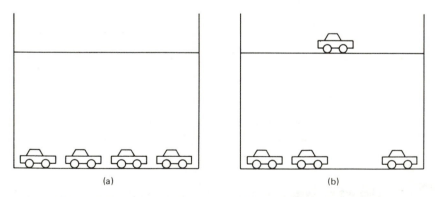

(a)   (b)

Figure 2.7. Simple "parking-station" analog of conduction processes in a semiconductor:
(a)   No movement possible.
(b)   Movement possible on both upper and lower levels.

in Fig. 2.7(b), the car on the second level is free to move as much as desired. This corresponds to an electron in the conduction band in a semiconductor. There will now be a vacant position on the lower level. Cars adjacent to this position can move into it, leaving a new vacant position behind. Hence, car motion is now possible on the lower level as well. This motion corresponds to the motion of electrons in the valence band. Instead of regarding the motion on the lower level as the result of the movements of a number of cars, it can be more simply described as the motion of the single vacant position. Similarly, in a crystal it is easier to think in terms of the motion of vacant states in the valence band. In many situations the correct motion of the vacancy can be predicted if it is regarded as a physical particle of positive charge commonly called a *hole*. Hence, current flow in a semiconductor can be regarded as being due to the sum of the motion of electrons in the conduction band and holes in the valence band.

2.6 DYNAMICS OF ELECTRONS AND HOLES

The motion of electrons and holes in semiconductors in response to applied forces differs from that of particles in free space. In addition to the applied force, there is always the effect of the periodic forces of the crystal atoms. However, results of quantum mechanical calculations indicate that, in most situations of interest herein, concepts developed for particles in free space can be applied to electrons and holes in semiconductors, with some modifications.

For example, in the case of electrons in a crystal lying within the conduction band, Newton's law becomes

$$F = m_e^* a = \frac{dp}{dt} \qquad (2.2)$$

where F is the applied force, m_e^* an "effective mass" of the electron which incorporates the effect of the periodic force of the lattice atoms, and p is known as the crystal momentum analogous to free-space momentum.

For a free electron, energy and momentum are related by a parabolic law,

$$E = \frac{p^2}{2m} \qquad (2.3)$$

For carriers in semiconductors, the situation can be more complex. In some semiconductors, an analogous law holds for electrons in the conduction band at energies close to the minimum, E_c, in this band:

$$E - E_c = \frac{p^2}{2m_e^*} \qquad (2.4)$$

A similar relationship holds for holes near the maximum energy, E_v, in the valence band:

$$E_v - E = \frac{p^2}{2m_h^*} \qquad (2.5)$$

The foregoing relationships are indicated in Fig. 2.8. Such semiconductors are known as *direct-band-gap semiconductors*, and the most important technologically is the compound semiconductor GaAs.

In other semiconductors, the minimum of the conduction band can be at a finite value of crystal momentum, obeying a relationship:

$$E - E_c = \frac{(p - p_0)^2}{2m_e^*} \qquad (2.6)$$

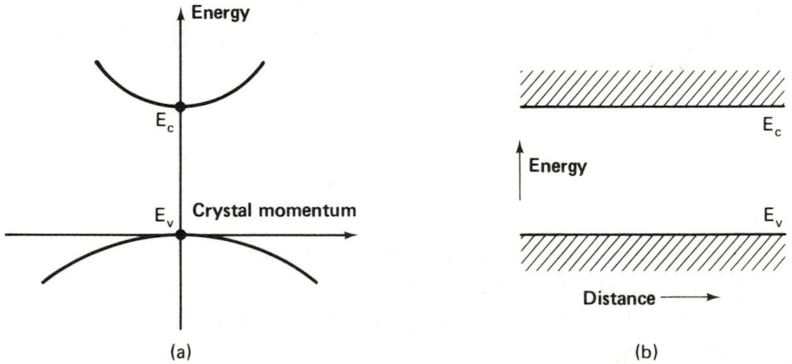

(a)   (b)

Figure 2.8. (a) Energy–crystal momentum relationships near the band edges for electrons in the conduction band and holes in the valence band of a direct-band-gap semiconductor. (b) Corresponding spatial representation of allowed energies in a semiconductor.

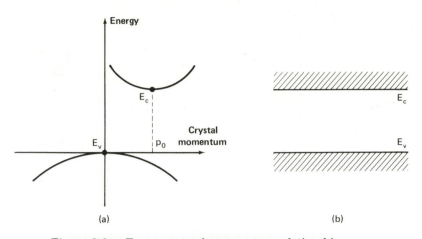

Figure 2.9. Energy–crystal momentum relationships near the band edges for an indirect-band-gap semiconductor. Also shown is the energy band spatial representation.

The valence band can exhibit a similar relationship:

$$E_v - E = \frac{(p - p_0')^2}{2m_h^*} \tag{2.7}$$

If $p_0 = p_0'$, the semiconductor has a direct band-gap. However, if $p_0 \neq p_0'$, the band-gap is called *indirect*. The most common elemental semiconductors, Ge and Si, both are indirect-band-gap materials. In each case, $p_0' = 0$ but p_0 is finite. Such a situation is shown in Fig. 2.9.

Note that the common representation of energy relations in semiconductor devices where energy is plotted as a function of distance (as also indicated in Figs. 2.8 and 2.9) does not differentiate between direct- and indirect-band-gap semiconductors.

2.7 ENERGY DENSITY OF ALLOWED STATES

The number of allowed states per unit volume in a semiconductor is obviously zero for energies corresponding to the forbidden gap and nonzero in the allowed bands. The question arises as to just how many states for electrons are distributed within the allowed bands.

An answer can be found reasonably simply (Refs. 2.1 to 2.4), at least for energies near the edges of the allowed bands, where car-

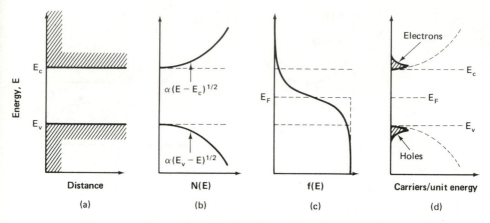

Figure 2.10. (a) Band representation of a semiconductor. (b) Corresponding energy density of allowed states for electrons. (c) Probability of occupation of these states. (d) Resulting energy distribution of electrons and holes. Note that most are clustered near the edge of the respective band.

riers can be treated similarly to free carriers. The number of allowed states per unit volume and energy, $N(E)$, at an energy E near the conduction-band edge (in the absence of anisotropy) is given by

$$N(E) = \frac{8\sqrt{2}\,\pi m_e^{*3/2}}{h^3}\ (E - E_c)^{1/2} \qquad (2.8)$$

where h is Planck's constant. A similar expression holds for energies near the valence-band edge. These distributions of allowed states are shown in Fig. 2.10(b).

2.8 DENSITIES OF ELECTRONS AND HOLES

Knowing the density of allowed states [Eq. (2.8)] and the probability of occupation of these states [Eq. (2.1)], it is now possible to calculate the actual energy distribution of electrons and holes. The results are shown schematically in Fig. 2.10.

Owing to the nature of the Fermi–Dirac distribution function, most of the electrons in the conduction band and holes in the valence band are clustered near the band edges. The total number in either band can be found by performing an integration. The number of electrons in the conduction band per unit volume of the crystal,

24

n, is given by

$$n = \int_{E_c}^{E_c \text{ max}} f(E) \, N(E) \, dE \qquad (2.9)$$

Since E_c is many kT larger than E_F, $f(E)$ for the conduction band reduces to

$$f(E) \approx e^{-(E - E_F)/kT} \qquad (2.10)$$

and the upper limit, $E_{c \text{ max}}$, can be replaced by infinity with little error. Therefore,

$$n = \int_{E_c}^{\infty} \frac{8\sqrt{2} \, \pi m_e^{*3/2}}{h^3} \, (E - E_c)^{1/2} \, e^{(E_F - E)/kT} \, dE$$

$$= \frac{8\sqrt{2} \, \pi}{h^3} \, m_e^{*3/2} \, e^{E_F/kT} \int_{E_c}^{\infty} (E - E_c)^{1/2} \, e^{-E/kT} \, dE \qquad (2.11)$$

Changing the variable of integration to $x = (E - E_c)/kT$ gives

$$n = \frac{8\sqrt{2} \, \pi}{h^3} \, (m_e^* kT)^{3/2} \, e^{(E_F - E_c)/kT} \int_0^{\infty} x^{1/2} \, e^{-x} \, dx \qquad (2.12)$$

The integral in this expression is in standard form and equals $\sqrt{\pi}/2$. Hence,

$$n = 2 \left(\frac{2\pi m_e^* kT}{h^2} \right)^{3/2} e^{(E_F - E_c)/kT} \qquad (2.13)$$

$$\boxed{n = N_c \, e^{(E_F - E_c)/kT}} \qquad (2.14)$$

where N_c is a constant at fixed T known as the *effective density of states in the conduction band* and is defined by comparing Eqs. (2.13) and (2.14). Similarly, the total number of holes in the valence band per unit volume of the crystal is given by

$$\boxed{p = N_V \, e^{(E_v - E_F)/kT}} \qquad (2.15)$$

with N_V, the effective density of states in the valence band, similarly defined.

For the idealized case of a pure and perfect semiconductor without surfaces, n equals p because each electron in the conduction band leaves a vacancy or hole in the valence band. Hence,

$$n = p = n_i \tag{2.16}$$

$$np = n_i^2 = N_C N_V e^{(E_v - E_c)/kT}$$
$$= N_C N_V e^{-E_g/kT} \tag{2.17}$$

where n_i is known as the "intrinsic concentration" and E_g is the width of the forbidden gap between the conduction and valence bands. Note also from Eq. (2.16) that

$$N_C e^{(E_F - E_c)/kT} = N_V e^{(E_v - E_F)/kT} \tag{2.18}$$

which gives

$$E_F = \frac{E_c + E_v}{2} + \frac{kT}{2} \ln \left(\frac{N_V}{N_C} \right) \tag{2.19}$$

Hence, the Fermi level in a pure and perfect semiconductor lies close to midgap, being offset only by differences in the effective density of states in the conduction and valence bands.

2.9 BOND MODEL OF A GROUP IV SEMICONDUCTOR

Some of the more fundamental of the semiconductor properties described up to now can be looked at from a different viewpoint for a class of semiconductors represented by those belonging to group IV of the periodic table of chemical elements. Although the following "bond model" description is not universally valid for all semiconducting material, it does allow the effects of impurities upon the electronic properties of semiconductors to be introduced in a simple manner.

The characteristic lattice structure of a semiconductor from group IV of the periodic table was shown in Fig. 2.3. A schematic two-dimensional representation of the silicon lattice is shown in Fig. 2.11(a). Each silicon atom is bonded to four neighbors by covalent bonds. Each covalent bond requires two electrons. Silicon has four valence electrons, so each covalent bond shares an electron originating from the central atom and one originating from the neighboring atom.

For the case shown in Fig. 2.11(a), the semiconductor cannot conduct electricity. However, at higher temperatures, some electrons in the covalent bond can obtain enough energy to break free from the bond as shown in Fig. 2.11(b). In this case, the electrons released are free to move about throughout the crystal and can contribute to current flows. Electrons in covalent bonds in the vicinity of the broken bond can also move into the location left vacant in this bond, leaving another broken bond behind. This process also contributes to current flow.

Reverting to the terminology of previous sections, an electron released from a covalent bond can be recognized as being in the conduction band, whereas those associated with covalent bonds are in the valence band. A broken bond can be identified as a hole in the valence band. The minimum energy required to release an electron from a covalent bond is then equal to the width of the forbidden band gap in the semiconductor.

The bond model is particularly useful for discussing the effects of impurities in silicon upon its electronic properties. In the

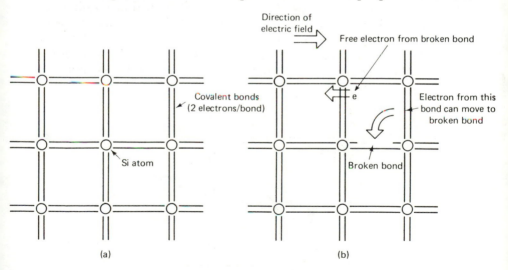

Figure 2.11. Schematic representation of the silicon crystal lattice.

(a)   No covalent bonds broken.

(b)   One covalent bond broken, showing the motion of the released electron as well as the motion of a nearby bonded electron into the position left vacant.

next section, the effects of very specialized impurities known as *dopants* are described.

2.10 GROUP III AND V DOPANTS

Impurity atoms can be incorporated in a crystal structure in two ways. They can occupy positions squeezed in between the atoms of the host crystal, in which case they are known as *interstitial impurities*. Alternatively they can substitute for an atom of the host crystal, maintaining the regular atomic arrangement in the crystal structure, in which case they are known as *substitutional impurities*.

Atoms from groups III and V of the periodic table act as substitutional impurities in silicon. A portion of the lattice where a group V impurity (e.g., phosphorus) has replaced a silicon atom is shown in Fig. 2.12. Four of the valence electrons are used up in covalent bonds, but the fifth is in a different situation. It is not in a covalent bond, so it is not in the valence band. For the case shown, it is tied to the group V atom and so is not free to move through the lattice. Hence, it is not in the conduction band either.

It might be expected that only a small amount of energy is required to release this extra electron compared to that required to free electrons locked up in covalent bonds. This is in fact the case. A rough estimate of the energy required can be found by noting the similarity to an electron tied to a hydrogen atom. The expression for the ionization energy (the energy required to release the electron) in

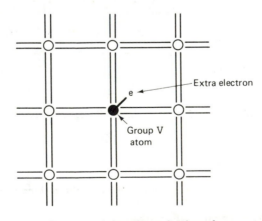

Figure 2.12. Portion of the silicon lattice where a group V atom has replaced a silicon atom.

the latter case is (Refs. 2.1 to 2.4)

$$E_i = \frac{m_0 q^4}{8\epsilon_0^2 h^2} = 13.6 \text{ electron volts (eV)} \qquad (2.20)$$

where m_0 is the electron's rest mass, q the electronic charge, and ϵ_0 the permittivity of free space. The extra electron orbits the group V atom, which has one unneutralized positive charge. The expression for the ionization energy in this case is therefore similar. The radius of the orbit turns out to be much larger than the interatomic distance, so ϵ_0 in Eq. (2.20) should be replaced by the permittivity of silicon (11.7ϵ_0). Since the orbiting electron experiences the periodic forces of the silicon lattice, the electron's mass also should be replaced by an effective mass ($m_e^*/m_0 \simeq 0.2$ for silicon). Hence, the energy required to free the extra electron is given by

$$E_i' \approx \frac{13.6(0.2)}{(11.7)^2} \approx 0.02 \text{ eV} \qquad (2.21)$$

This is much less than the band-gap energy of silicon of 1.1 eV. A free electron is in the conduction band. Hence, the extra electron tied to the group V atom lies at an energy E_i' below the edge of the conduction band, as illustrated in Fig. 2.13(a). Note that this places an allowed energy level within the "forbidden" gap.

In an analogous way, a group III impurity does not have enough valence electrons to satisfy the four covalent bonds. This gives rise to a hole tied to the group III atom. The energy required to release the hole is similar to that given by Eq. (2.21). Hence, a group III atom gives rise to an allowed energy level for electrons in the forbidden gap just above the valence-band edge, as shown in Fig. 2.13(b).

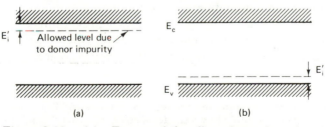

(a)   (b)

Figure 2.13.   (a)   Energy of the allowed state introduced into the forbidden gap by a group V substitutional impurity. (b)   Corresponding energy state for a group III impurity.

2.11 CARRIER DENSITIES

Since the energy required to release the extra electron from a group V atom is small, it is not unexpected that, at room temperature, most of these electrons have acquired this energy. Hence, most have left the group V atom, with its net positive charge, behind and are free to move through the crystal. Since group V atoms donate electrons to the conduction band, they are known as *donors*. A more quantitative idea of the number of electrons that have obtained the small amount of energy required can be obtained by referring to Fig. 2.14. Note that the form of the Fermi–Dirac distribution function indicates that donor levels have only a small probability of being occupied.[1] This means that most electrons have left the donor site and are in the conduction band.

The total number of electrons in the conduction band and holes in the valence band in this case can be found by considering the condition for charge neutrality in the semiconductor:

$$p - n + N_D^+ = 0 \qquad (2.22)$$

where p is the hole density in the valence band, n the density of conduction band electrons, and N_D^+ the density of ionized donors (i.e., positive charges left behind when the electron departs). The other important equation comes from Eq. (2.17):

$$np = n_i^2 \qquad (2.23)$$

This relation is more general than for the case of pure semiconductors previously discussed. Equations (2.14), (2.15), and (2.22), in conjunction with the Fermi–Dirac distribution function, can be solved to give precise values for n, p, and N_D^+ under general conditions. However, for most cases of interest in this book, the approximate but much simpler method of solution outlined below will give results of more than adequate accuracy.

Since the vast majority of donors will be ionized, N_D^+ will be nearly equal to the total density of donors, N_D. From Eq. (2.22), n will be greater than p and, in fact, very much greater when N_D be-

---

[1] The statistics governing the occupation of the donor level are actually slightly different from those governing the occupation of levels within the allowed band. Once a donor level is occupied by a single electron of either "spin," the effective positive charge on the central donor atom is neutralized and there is no attraction that would allow occupation by a second electron of opposite spin. This results in an expression for the probability of occupation which differs slightly from the Fermi–Dirac function. That difference is not important in this book.

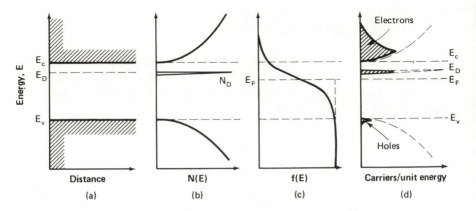

Figure 2.14. (a) Band representation of a group IV semi-conductor with a group V substitutional impurity of density N_D per unit volume. (b) Corresponding energy density of allowed states. (c) Probability of occupation of these states. (d) Resulting energy distributions of electrons and holes. (The case shown would correspond to quite high temperatures. At more moderate temperatures the probability of occupation of donor states by electrons would be even smaller than shown.)

comes large. Hence, the approximate solution is

$$N_D^+ \approx N_D$$

$$n \approx N_D$$ (2.24)

$$p \approx \frac{n_i^2}{N_D} \ll n$$

An analogous situation occurs with group III impurities. These very easily give up their excess hole to the valence band or equivalently accept an electron from this band. Consequently, they are known as *acceptors*. An ionized acceptor has a net negative charge. Hence,

$$p - n - N_A^- = 0$$ (2.25)

where N_A^- is the density of ionized acceptors.

The approximate solution in this case is

$$N_A^- \approx N_A$$

$$p \approx N_A$$ (2.26)

$$n \approx \frac{n_i^2}{N_A} \ll p$$

31

2.12 LOCATION OF FERMI LEVEL IN DOPED SEMICONDUCTORS

The equations for electron and hole densities derived in Eqs. (2.14) and (2.15) apply to more general cases than that for pure semiconductors. For the case of material doped with donors (commonly called *n-type material*), these become

$$n = N_D = N_C e^{(E_F - E_c)/kT} \tag{2.27}$$

or, equivalently,

$$E_F - E_c = kT \ln \left(\frac{N_D}{N_C} \right) \tag{2.28}$$

Similarly, for material doped with acceptors (*p-type material*),

$$p = N_A = N_V e^{(E_v - E_F)/kT} \tag{2.29}$$

$$E_v - E_F = kT \ln \left(\frac{N_A}{N_V} \right) \tag{2.30}$$

As the semiconductor material becomes more heavily doped, the Fermi level E_F moves away from midgap and approaches the conduction band for *n*-type material or the valence band for *p*-type material, as shown in Fig. 2.15.

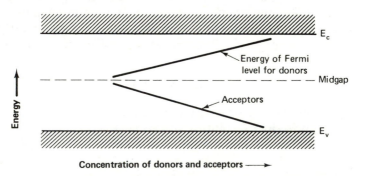

Figure 2.15. Energy of the Fermi level as a function of the concentration of donors and acceptors.

2.13   EFFECT OF OTHER TYPES OF IMPURITIES

Our theoretical understanding of the properties of impurities in silicon other than those from groups III and V is less developed, although the practical effects of such impurities are well known.

Just as group III and V impurities introduce an allowed energy level into the forbidden band gap of silicon, so do more general impurities. This is indicated in Fig. 2.16, which shows the al-

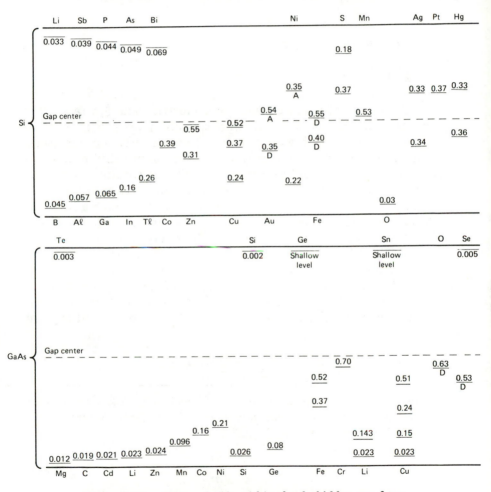

Figure 2.16.   Energy levels within the forbidden gap for a range of impurities in Si and GaAs. A indicates an acceptor level, D a donor level. [After S. M. Sze and J. Irwin, *Solid-State Electronics 11* (1968), 599.]

lowed energy levels introduced by a range of impurities in the silicon as well as the compound semiconductor GaAs. Some impurities introduce multiple energy levels, as shown. Crystal defects act in a similar way to introduce allowed levels into the forbidden gap.

Impurities, particularly those which introduce energy levels near the middle of the band gap, generally degrade the properties of semiconductor devices. Impurity concentrations in the starting material used in the fabrication of these devices are therefore kept as low as technology will allow—generally less than 1 part per billion.

2.14 CARRIER TRANSPORT

2.14.1 Drift

Under the influence of an applied electric field, ξ, a randomly moving free electron would have an acceleration $a = \xi/m$ in a direction opposite to the field, with its velocity in this direction increasing with time. The electron in a crystal structure is in a different situation. It moves with a different mass and will not continue accelerating for very long. It will eventually collide with a lattice atom, or an impurity atom, or a defect in the crystal structure. Such a collision will tend to randomize the electron's motion. In other words, it will tend to reduce the excess velocity that the electron picked up in the applied field. The "average" time between collisions is called the *relaxation time*, t_r. This will be determined by the random thermal velocity of electrons, which is generally much larger than field imparted velocities. The average velocity increase of electrons between collisions caused by the field is called the *drift velocity* and is given by

$$v_d = \frac{1}{2}at = \frac{1}{2}\frac{qt_r}{m_e^*}\xi \qquad (2.31)$$

for electrons in the conduction band. (The factor of 2 disappears if t_r is averaged over all electron velocities.) The electron carrier mobility is defined by the ratio

$$\mu_e = \frac{v_d}{\xi} = \frac{qt_r}{m_e^*} \qquad (2.32)$$

The corresponding current density flow due to conduction band electrons will be

$$J_e = qnv_d = q\mu_e n\xi \qquad (2.33)$$

An analogous equation for holes in the valence band is

$$J_h = q\mu_h p\xi \tag{2.34}$$

The total current flow is just the sum of these two components. Hence, the conductivity, σ of the semiconductor can be identified as

$$\sigma = \frac{1}{\rho} = \frac{J}{\xi} = q\mu_e n + q\mu_h p \tag{2.35}$$

where ρ is the resistivity.

Although the analysis resulting in Eq. (2.32) is simplistic, it does allow an intuitive understanding of how the carrier mobilities, μ_n and μ_p, change with changes in the density of dopants, temperature, and electric field strength.

For relatively pure semiconductors of good crystallographic quality, the collisions that randomize the carrier velocities will involve the atoms of the host crystal. However, ionized dopants are very effective scatterers because of their associated net charge. Consequently, as the semiconductor becomes more heavily doped, the average time between collisions and hence the mobility will decrease. For good-quality silicon, empirical expressions relating the carrier mobilities to the level of dopants N (in cm^{-3}) are (Ref. 2.5)

$$\mu_e = 65 + \frac{1265}{1 + (N/8.5 \times 10^{16})^{0.72}} \; cm^2/V\text{-}s$$
$$\mu_h = 47.7 + \frac{447.3}{1 + (N/6.3 \times 10^{16})^{0.76}} \; cm^2/V\text{-}s \tag{2.36}$$

Less specialized impurities as well as crystal defects will decrease mobilities further, for similar reasons.

Increasing temperature will increase the vibration of the host atoms, making them larger "targets," again decreasing the average time between collisions as well as the carrier mobilities. This effect becomes less pronounced at high doping levels, where ionized dopants are effective carrier scatterers.

Increasing the strength of the electric field will eventually increase the drift velocities of carriers to values where they will become comparable to the random thermal velocities. Hence, the total velocity of electrons will ultimately increase with field strength, decreasing the time between collisions and the mobility.

2.14.2 Diffusion

Apart from motion by drift, carriers in semiconductors can also flow by diffusion. It is a well-known physical effect that any excess concentration of particles such as gas molecules will tend to dissipate itself unless constrained. The basic cause of this effect is the random thermal velocity of the particles involved.

The flux of particles is proportional to the negative of the concentration gradient (Fig. 2.17). Since current is proportional to the flux of charged particles, the current density corresponding to a one-dimensional concentration gradient of electrons is

$$J_e = qD_e \frac{dn}{dx} \qquad (2.37)$$

where D_e is a constant known as the diffusion constant. Similarly, for holes

$$J_h = - qD_h \frac{dp}{dx} \qquad (2.38)$$

Note the sign difference between Eqs. (2.37) and (2.38), which is due to the opposite types of charges involved. Drift and diffusion processes are fundamentally related and the mobilities and diffusion constants are not independent. They are interconnected by the *Einstein relations*

$$D_e = \frac{kT}{q}\mu_e \qquad \text{and} \qquad D_h = \frac{kT}{q}\mu_h \qquad (2.39)$$

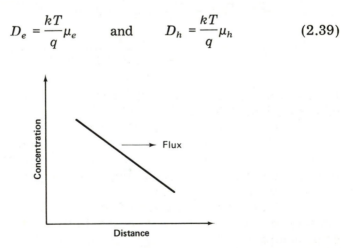

Figure 2.17.   Diffusive flux of carriers in the presence of a concentrator gradient.

kT/q is a parameter that will appear often in relation to solar cells. It has the dimensions of voltage and the value of 26 mV at room temperature, a value worth committing to memory!

2.15 SUMMARY

The major points arising from this chapter are the following. Semiconductors have an electronic structure such that one band of allowed states virtually completely occupied by electrons (the valence band) is separated by a forbidden energy gap from the next band of allowed states, which is virtually devoid of electrons (the conduction band). Current flow in semiconductors is due to both motion of electrons in the conduction band and the effective motion of vacancies or holes in the valence band. In many situations, electrons in the conduction band and holes in the valence band can be regarded as free particles, provided that an "effective" mass is used to include the effect of periodic forces of the host atoms in the crystal. Most conduction-band electrons have energies close to that of the conduction-band edge, whereas most holes have energies close to that of the valence-band edge.

Semiconductors can be divided into "direct"- and "indirect"-band-gap types, depending on the form of the relationship between the energy of electrons in the conduction band and their crystal momentum.

Specialized impurities known as dopants, when introduced into semiconductors, can control the relative concentrations of electrons in the conduction band of a semiconductor and holes in the valence band. Carriers in these bands can flow by drift and diffusion when the appropriate perturbations are present.

In Chapter 3, additional electronic processes occurring within semiconductors when disturbed by light are described. From the fundamental mechanisms discussed in this and the next chapter, a single system of self-consistent equations will be synthesized. This system will be used in later chapters to establish the principles of solar cell design.

EXERCISES

2.1. For a crystal with a cubic unit cell, indicate on a sketch of the cell the following crystal planes: (a) (100); (b) (010); (c) (110); (d) (111).

2.2. (a) Silicon solar cell performance can be improved by selectively etching the cell surface to reduce reflection losses. Figure 7.6 shows a silicon crystal surface originally orientated parallel to the (100) plane which has been attacked by a chemical etch that etches at different rates in different directions through the crystal. This exposes the square-based pyramids shown. Given that the sides of the pyramids are all members of the {111} equivalent set of planes, find the angle between opposite faces of the pyramids.

(b) A fraction R of light incident normally on the original silicon surface was reflected. Neglecting dependencies upon angle of incidence and wavelength, show that fraction reflected after the selective etch is reduced to *slightly less* than R^2.

2.3. One method of introducing impurities into silicon in controlled quantities is a technique known as *ion implantation*. Ions of the desired impurity are accelerated to high velocity and directed at the silicon surface. If the ions impinged parallel to each of the crystal directions shown in Fig. 2.3(b)-(d), in which case would you expect the ions to penetrate the greatest distance into the silicon?

2.4. An allowed state for an electron in a semiconductor lies at an energy equal to 0.4 eV above the Fermi level. What probability has this state of being occupied by an electron under thermal equilibrium conditions at 300 K?

2.5. Assuming that the effective masses of electrons and holes are equal to the free electron mass, calculate the effective density of states in the conduction and valence bands for silicon at 300 K. Assuming a band gap of 1.1 eV, find the intrinsic concentration in silicon at this temperature.

2.6. (a) Silicon is uniformly doped with 10^{22} phosphorus atoms/m^3. Assuming that all these donor impurities are ionized, estimate the concentration of electrons and holes in this material under thermal equilibrium at 300 K. Hence, calculate the energy of the Fermi level in this material below the conduction-band edge.

(b) Given that the donor level for phosphorus lies 0.045 eV below the conduction-band edge, calculate the probability that this level is occupied by an electron and hence check on the assumption that all donors are ionized. (Use $N_C = 3 \times 10^{25}$ m^{-3}, $N_V = 10^{25}$ m^{-3}, and $n_i = 1.5 \times 10^{16}$ m^{-3}.)

2.7. Using Eq. (2.36) for the electron and hole mobilities in silicon, estimate the resistivity of the silicon specimen of Exercise 2.6.

2.8. Estimate the average time between collisions with the host atoms for electrons in the conduction band of lightly doped silicon.

2.9. An electric field of 10^4 V/m is applied to a specimen of silicon at 300 K doped with 10^{22} donors/m^3. Given that the thermal velocity is 10^5 m/s, compare the drift and thermal velocities for conduction-band electrons. At what value of the field strength will these be comparable?

2.10. In a section of silicon at 300 K, the field strength is zero and conduction band electrons have a concentration that varies from 10^{22} per m^2

to 10^{21} per m^2 over a distance of 1 μm. Assuming a linear variation of electrons, calculate the corresponding current density.

REFERENCES

[2.1] V. AZAROFF AND J. J. BROPHY, *Electronic Processes in Materials* (New York: McGraw-Hill, 1963).

[2.2] A. VAN DER ZIEL, *Solid State Physical Electronics*, 3rd ed. (Englewood Cliffs, N.J.: Prentice-Hall, 1976).

[2.3] S. WANG, *Solid-State Electronics* (New York: McGraw-Hill, 1966).

[2.4] W. SHOCKLEY, *Electrons and Holes in Semiconductors* (New York: Van Nostrand Rheinhold, 1950).

[2.5] D. M. CAUGHEY AND R. E. THOMAS, "Carrier Mobilities in Silicon Empirically Related to Doping and Field," *Proceedings of the IEEE 55* (1967), 2192–2193.

Chapter

3

GENERATION, RECOMBINATION, AND THE BASIC EQUATIONS OF DEVICE PHYSICS

3.1 INTRODUCTION

In Chapters 1 and 2, the relevant properties of sunlight and semiconductors have been reviewed. In the present chapter, the interaction between these two basic components of a solar photovoltaic system is examined.

This leads to the concepts of generation and recombination of excess carriers within the semiconductor material and a description of the physical mechanisms involved. Finally, the material discussed in connection with semiconductor properties is combined into a basic set of equations capable of describing the ideal properties of most semiconductor devices, including solar cells.

3.2 INTERACTION OF LIGHT WITH SEMICONDUCTOR

Figure 3.1 shows a ray of monochromatic light incident perpendicularly to a flat section of semiconductor. A certain fraction of the

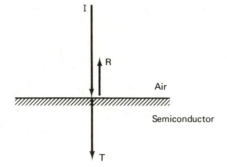

Figure 3.1. Ray of monochromatic light incident on a semiconductor.

incident power, R, will be reflected and the remainder, T, transmitted into the semiconductor.

The transmitted light can be absorbed within the semiconductor by using its energy to excite electrons from occupied low-energy states to unoccupied higher-energy states. Since there are a large number of occupied states within the valence band of a semiconductor separated by the forbidden band from largely unoccupied states in the conduction band, absorption is particularly likely when the energy of the photons making up the light is larger than the forbidden band gap, E_g, of the semiconductor.

An absorbing material has an *index of refraction*, $\hat{n}_c$, which is a complex number. This index can be written as $\hat{n}_c = \hat{n} - i\hat{k}$, where $\hat{k}$ is known as the *extinction coefficient*. The two components of this complex number for silicon are shown in Fig. 3.2 as a function of the wavelength of the incident light. The fraction of light reflected for normal incidence is given by (Refs. 3.1 and 3.2)

$$R = \frac{(\hat{n} - 1)^2 + \hat{k}^2}{(\hat{n} + 1)^2 + \hat{k}^2} \qquad (3.1)$$

Substituting the appropriate values for silicon shows that for all wavelengths of interest in solar cell work, over 30% of the incident light is reflected. This is clearly undesirable from the point of view of making efficient solar cells. Antireflection coatings and other techniques (see Exercise 2.2) are used in solar cells to reduce this figure as much as possible.

The transmitted light is attenuated as it passes through the semiconductor. The rate of absorption of light is proportional to the intensity (the flux of photons) for a given wavelength. This common

41

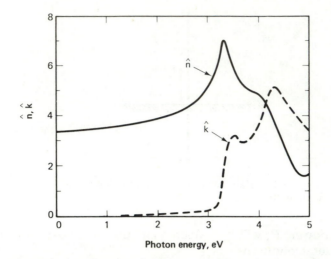

Figure 3.2. Real and (negative) imaginary parts of the refractive index for silicon. [After H. R. Phillip and E. A.
Taft, *Physical Review 120* (1960), 37–38.]

physical occurrence leads to an exponential decay in intensity of monochromatic light as it passes through the semiconductor, described
mathematically as

$$\mathcal{I}(x) = \mathcal{I}(x_0)e^{-\alpha(x-x_0)} \tag{3.2}$$

where α is a function of wavelength and is known as the *absorption
coefficient*. This parameter is important in solar cell design because
it determines how far below the surface of the cell light of a given
wavelength is absorbed.

The absorption coefficient α and the extinction coefficient $\hat{k}$
are not unrelated. Describing the light by a plane wave of frequency
f propagating in the x direction with velocity v, the associated electric
field strength is (Ref. 3.2)

$$\xi = \xi_0 \exp\left[i2\pi f\left(t - \left(\frac{x}{v}\right)\right)\right] \tag{3.3}$$

The velocity in the semiconductor is related to the velocity in
vacuum, c, by

$$v = \frac{c}{\hat{n}_c} \tag{3.4}$$

Hence,

$$\frac{1}{v} = \frac{\hat{n}}{c} - \frac{i\hat{k}}{c} \tag{3.5}$$

Substituting into Eq. (3.3) gives

$$\xi = \xi_0 \exp\left(i2\pi ft\right) \exp\left(-\frac{i2\pi x \hat{n}}{c}\right) \exp\left(-\frac{2\pi f \hat{k} x}{c}\right) \tag{3.6}$$

The last term in Eq. (3.6) is an attenuation factor. Power will attenuate as the square of the electric field strength. Comparing Eqs. (3.2) and (3.6) gives the relationship

$$\alpha = \frac{4\pi f \hat{k}}{c} \tag{3.7}$$

3.3 ABSORPTION OF LIGHT

3.3.1 Direct-Band-Gap Semiconductor

Fundamental absorption refers to the annihilation or absorption of photons by the excitation of an electron from the valence band up into the conduction band, leaving a hole in the valence band. Both energy and momentum must be conserved in such a transition. A photon has quite a large energy (hf) but a small momentum (h/λ).

The form of the absorption process for a direct-band-gap semiconductor is shown in the energy–momentum sketch of Fig. 3.3. Because the photon momentum is small compared to the crystal momentum, the latter essentially is conserved in the transition. The energy difference between the initial and final state is equal to the energy of the original photon:

$$E_f - E_i = hf \tag{3.8}$$

In terms of the parabolic bands described in Chapter 2,

$$E_f - E_c = \frac{p^2}{2m_e^*}$$
$$E_i - E_v = \frac{p^2}{2m_h^*} \tag{3.9}$$

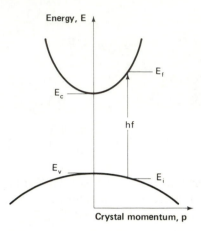

Figure 3.3. Energy–crystal momentum diagram of a direct-band-gap semiconductor, showing the absorption of a photon by the excitation of an electron from the valence to the conduction band.

Therefore, the specific value of crystal momentum at which the transition occurs is given by

$$hf - E_g = \frac{p^2}{2}\left(\frac{1}{m_e^*} + \frac{1}{m_h^*}\right) \qquad (3.10)$$

As the photon energy hf increases, so does the value of crystal momentum at which the transition occurs (Fig. 3.3). The energy away from the band edge of both the initial and final states also increases. The probability of absorption depends on the density of electrons at the energy corresponding to the initial state as well as the density of empty states at the final energy. Since both these quantities increase with energy away from the band edge, it is not surprising that the absorption coefficient increases rapidly with increasing photon energy above E_g. A simple theoretical treatment gives the result (Ref. 3.2)

$$\alpha(hf) \approx A^*(hf - E_g)^{1/2} \qquad (3.11)$$

where A^* is a constant having the numerical value of 2×10^4 when α is expressed in cm^{-1} and hf and E_g are in electron volts (eV). A comparison of this form of expression and experimental results for GaAs, a direct-band-gap semiconductor, is shown in Fig. 3.4. There is reasonable agreement at the higher values of the absorption coefficient.

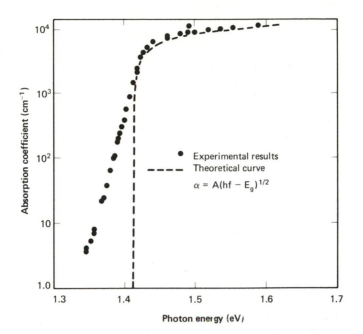

Figure 3.4. Absorption coefficient of GaAs as a function of photon energy. [After T. S. Moss and T. D. F. Hawkins, *Infrared Physics 1*, (1961), 111.]

Since the light intensity drops to $1/e$ of its value in passing a distance $1/\alpha$ through the semiconductor, Eq. (3.11) shows that sunlight of photon energy greater than E_g is absorbed within the first few microns of entering a direct-band-gap semiconductor.

3.3.2 Indirect-Band-Gap Semiconductor

In the case of an indirect-band-gap semiconductor, the minimum energy in the conduction band and the maximum energy in the valence band occur at different values of crystal momentum (Fig. 3.5). Photon energies much larger than the forbidden gap are required to give direct transitions of electrons from the valence to the conduction band, the process described in Section 3.3.1.

However, transitions can occur at lower energies by a two-step process involving not only photons and electrons but also a third particle, a *phonon*. In the same way as light can be thought of as either waves or particles, so can the coordinated vibration of the atoms making up the crystal structure. A phonon is just a *quantum* or fundamental particle corresponding to the coordinated vibration.

45

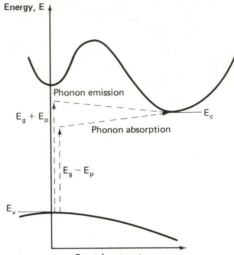

Figure 3.5. Energy–crystal momentum diagram of an indirect-band-gap semiconductor, showing the absorption of photons by two step processes involving phonon emission or absorption.

As opposed to photons, phonons have low energy but relatively high momentum.

This difference is explained by noting the relationship between phonons and sound, which propagates through a solid by coordinated atomic vibrations. The large difference between the velocity of light and sound in a solid is associated with the difference between the ratio of energy and momentum for the corresponding fundamental particles.

As indicated in the energy–momentum sketch of Fig. 3.5, an electron can make a transition from the maximum energy in the valence band to the minimum energy in the valence band in the presence of photons of suitable energy by the emission or absorption of a phonon of the required momentum. Hence, the minimum photon energy required to excite an electron from the valence to the conduction band is

$$hf = E_g - E_p \tag{3.12}$$

where E_p is the energy of an absorbed phonon with the required momentum.

Since the indirect-band-gap absorption process requires that an extra "particle" be involved, the probability of light being absorbed by this process is much less than in the direct-band-gap case. Hence,

46

the absorption coefficient is low and light can pass a reasonable distance into the semiconductor prior to absorption. An analysis of the theoretical value of the absorption coefficient gives the results (Ref. 3.2)

$$\alpha_a(hf) = \frac{A(hf - E_g + E_p)^2}{\exp(E_p/kT) - 1} \tag{3.13}$$

for a transition involving phonon absorption, and

$$\alpha_e(hf) = \frac{A(hf - E_g - E_p)^2}{1 - \exp(-E_p/kT)} \tag{3.14}$$

for one involving phonon emission. Since both phonon emission and absorption are possible for $hf > E_g + E_p$, the absorption coefficient is then

$$\alpha(hf) = \alpha_a(hf) + \alpha_e(hf) \tag{3.15}$$

The absorption coefficient for silicon is shown in Fig. 3.6 as a function of the wavelength of incident light at different temperatures. The weak absorption region at wavelengths greater than 0.5 μm corresponds to indirect-band-gap processes. Below 0.4 μm, the absorption coefficient increases very rapidly in a manner that could be attributed to direct gap absorption. An empirical expression involving terms of the form of Eqs. (3.11), (3.13), and (3.14) has been found to describe these experimental results to a high degree of accuracy over the photon energy range 1.1 to 4.0 eV and for the temperature range 20 to 500 K. This takes the form (Ref. 3.3)

$$\alpha(hf, T) = \sum_{\substack{i=1,2 \\ j=1,2}} A_{ij} \left\{ \frac{[hf - E_{gj}(T) + E_{pi}]^2}{\exp(E_{pi}/kT) - 1} + \frac{[hf - E_{gj}(T) - E_{pi}]^2}{1 - \exp(-E_{pi}/kT)} \right\}$$
$$+ A_d[hf - E_{gd}(T)]^{1/2} \tag{3.16}$$

where the values of the constants A_{ij}, E_{gj}, and E_{pi} are as given in Table 3.1.

3.3.3 Other Absorption Processes

Light absorption within a semiconductor is by no means limited to the processes discussed so far. It has already been indicated

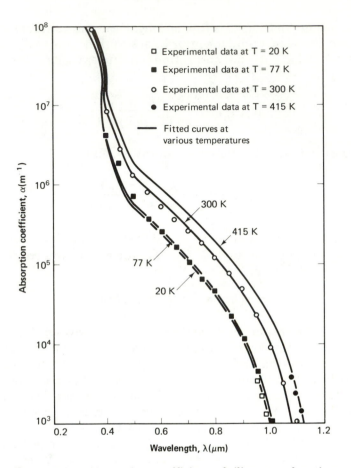

Figure 3.6. Absorption coefficient of silicon as a function of wavelength of light at different temperatures. (After Ref. 3.3.)

that absorption can occur at high-enough photon energy by excitation across the direct forbidden gap of an indirect-band-gap semiconductor such as silicon. Similarly, two-step absorption involving emission or absorption of phonons can also occur in direct-band-gap semiconductors, as indicated in Fig. 3.7(a). This occurs in parallel with the much stronger direct absorption process discussed in Section 3.3.1.

Similarly, a carrier can be excited to a higher energy in its respective band by a photon with the emission or absorption of a phonon as in Fig. 3.7(b). This process is relatively weak but, as might be expected, is strongest at long wavelengths when carrier concentrations are large. Although not important in solar cell work, it does

Table 3.1. VALUE OF CONSTANTS FOR THE EMPIRICAL
EXPRESSION FOR THE ABSORPTION COEFFICIENT
OF SILICON

Quantity	Numerical value
$E_{g1}(0)^*$	1.1557 eV
$E_{g2}(0)^*$	2.5 eV
$E_{gd}(0)^*$	3.2 eV
E_{p1}	1.827×10^{-2} eV
E_{p2}	5.773×10^{-2} eV
A_{11}	1.777×10^3 cm^{-1}/eV^2
A_{12}	3.980×10^4 cm^{-1}/eV^2
A_{21}	1.292×10^3 cm^{-1}/eV^2
A_{22}	2.895×10^4 cm^{-1}/eV^2
A_d	1.052×10^6 cm^{-1}/eV$^{1/2}$

$^*E_g(T) = E_g(0) - [\beta T^2/(T + \gamma)]$ with $\beta = 7.021 \times 10^{-4}$ eV/K   and $\gamma =$ 1108 K.
Source: Ref. 3.3.

demonstrate a case where absorption occurs without the generation of an electron–hole pair.

Light absorption can also occur by the excitation of transitions between the allowed bands in a semiconductor and energy levels introduced into the forbidden gap by impurities as shown in Fig. 3.8.

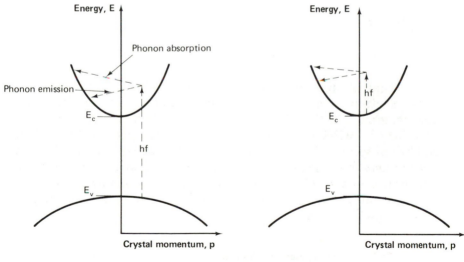

(a)   (b)

Figure 3.7. (a) Two-step photon absorption processes in direct-band-gap semiconductor. (b) Free carrier absorption in the conduction band which does not result in the generation of electron–hole pairs.

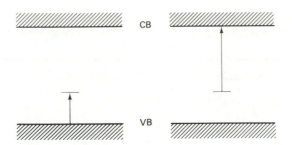

Figure 3.8.   Light absorption by carrier excitation between bands and energy levels within the forbidden gap.

Finally, two processes will be briefly mentioned which may produce second-order effects in solar cells. The Franz–Keldysh effect (Ref. 3.2) arises in the presence of a strong electric field such as that which occurs in some regions of solar cells. This effect shifts the absorption edge to lower energies, having the same effect as a reduction of the width of the forbidden band gap. High-dopant-impurity concentrations also affect the absorption edge. A reduction in the width of the forbidden gap again occurs in the presence of such concentrations.

3.4   RECOMBINATION PROCESSES

3.4.1   Relaxation to Equilibrium

Light of appropriate wavelength shining on a semiconductor creates electron–hole pairs. The concentrations of carriers in illuminated material will therefore be in excess of their values in the dark. If the light is switched off, these concentrations decay back to their equilibrium values. The process by which this decay occurs is known as *recombination*. Three different recombination mechanisms will be described in the following sections. These mechanisms can occur in parallel, in which case the recombination rate is just the sum of those for the individual processes.

3.4.2   Radiative Recombination

Radiative recombination is just the reverse of the absorption process described in Section 3.3. An electron occupying a higher-energy state than it would under thermal equilibrium makes a transition to an empty lower-energy state with all (or most) of the energy difference between states emitted as light. All the mechanisms considered for absorption have inverse radiative recombination processes (Fig. 3.9). Radiative recombination occurs more rapidly in direct-band-gap semiconductors than in indirect types because a two-step process involving a phonon is required for the latter.

The total radiative recombination rate, R_R, is proportional to the product of the concentration of occupied states (electrons) in the conduction band and that of unoccupied states in the valence band (holes):

$$R_R = Bnp \qquad (3.17)$$

where B is a constant for a given semiconductor. Because of the relationship between optical absorption and this recombination process, B can be calculated from the semiconductor's absorption coefficient (Ref. 3.2).

In thermal equilibrium when $np = n_i^2$, this recombination rate is balanced by an equal and opposite generation rate. In the absence of generation by external stimuli, the net recombination rate corresponding to Eq. (3.17), U_R, is given by the total recombination rate minus the equilibrium generation rate:

$$U_R = B(np - n_i^2) \qquad (3.18)$$

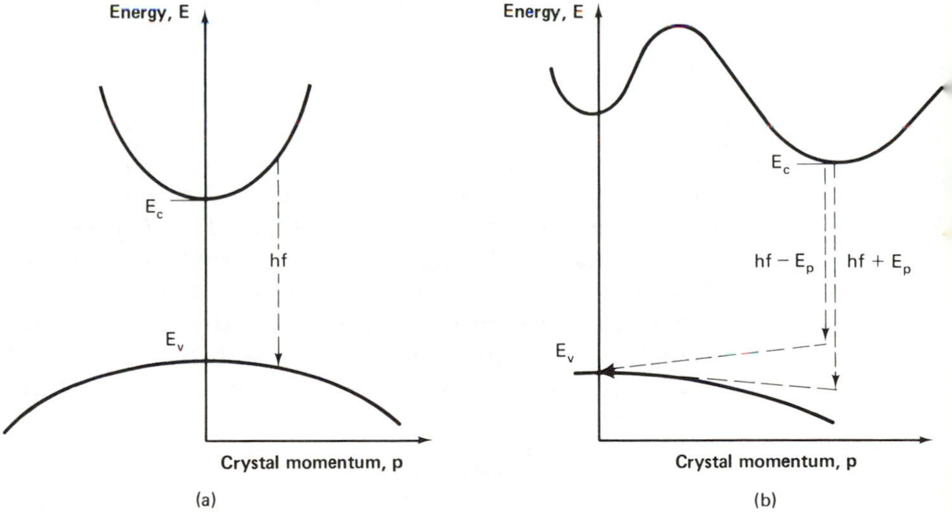

Figure 3.9.   Radiative recombination in semiconductors:
(a)   Direct band gap.
(b)   Indirect band gap.

With any recombination mechanism, it is possible to define associated *carrier lifetimes*, τ_e and τ_h, for electrons and holes:

$$\tau_e = \frac{\Delta n}{U}$$

$$\tau_h = \frac{\Delta p}{U} \tag{3.19}$$

where U is the net recombination rate and Δn and Δp are the disturbances of the respective carriers from their equilibrium values, n_0 and p_0.

For the radiative recombination mechanism with $\Delta n = \Delta p$, the characteristic lifetime determined from Eq. (3.18) is (Ref. 3.2)

$$\tau = \frac{n_0 p_0}{B n_i^2 (n_0 + p_0)} \tag{3.20}$$

For silicon, B has the value of about 2×10^{-15} cm^3/s (Ref. 3.2).

As might be expected, radiative recombination lifetimes are much smaller in direct-band-gap material than in indirect material. This process forms the basis of commercial semiconductor lasers and light-emitting diodes using GaAs and its alloys. For silicon, other recombination mechanisms are far more important.

3.4.3 Auger Recombination

In the Auger (pronounced something like "oh-shay") effect, the electron recombining with the hole gives the excess energy to a second electron (either in the conduction or valence band) instead of emitting light. This process is illustrated in Fig. 3.10. This second electron then relaxes back to its original energy by emitting phonons. Auger recombination is just the reverse process of the more familiar impact ionization effect, where a high-energy electron collides with an atom, breaking a bond and creating an electron–hole pair. The characteristic lifetime τ associated with the Auger process is (Ref. 3.2)

$$\frac{1}{\tau} = Cnp + Dn^2$$

or

$$\frac{1}{\tau} = Cnp + Dp^2 \tag{3.21}$$

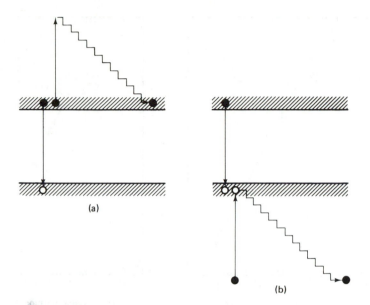

Figure 3.10. Auger recombination with associated excess energy given to an electron in:
(a) Conduction band.
(b) Valence band.

for materials with an abundance of electrons and holes, respectively. The first term on the right in each case describes electron excitation in the minority carrier band, and the second describes it in the majority carrier band. Auger recombination is particularly effective in relatively highly doped material due to this second term. For good-quality silicon, this is the dominant recombination process for impurity levels greater than 10^{17} cm^{-3}. The experimental results for the variation of lifetime with increasing dopant density for high-quality silicon shown in Fig. 3.11 indicate the rapid decrease at high dopant density due to Auger processes.

3.4.4   Recombination through Traps

It has been indicated in Chapter 2 that impurities and defects in semiconductors can give rise to allowed energy levels within the otherwise forbidden gap. These defect levels create a very efficient two-step recombination process whereby electrons relax from conduction-band energies to the defect level and then relax to the valence band, annihilating a hole as shown in Fig. 3.12(a).

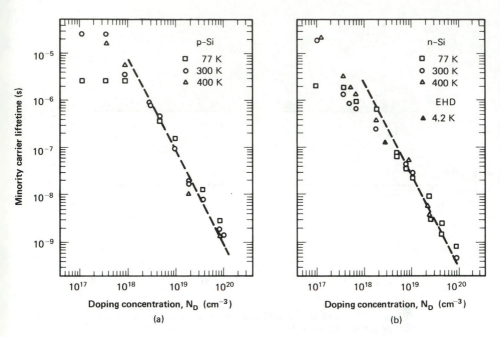

Figure 3.11. Experimental recombination lifetimes in high-quality silicon. The dashed lines represent the quadratic dependence predicted theoretically.

(a) p-type silicon
(b) n-type silicon
[After J. Dziewior and W. Schmid, *Applied Physics Letters 31* (1977), 346–348.]

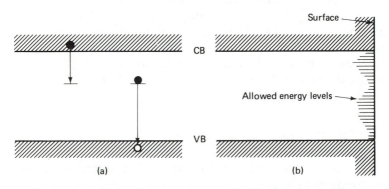

Figure 3.12. (a) Two-step recombination process via a trapping level within the forbidden gap of a semiconductor. (b) Surface states lying within the forbidden gap at the surface of a semiconductor.

54

The analysis of the dynamics of this process is straightforward but lengthy (Ref. 3.4). The result is that the net recombination-generation rate by traps, U_T, can be written

$$U_T = \frac{np - n_i^2}{\tau_{h0}(n + n_1) + \tau_{e0}(p + p_1)} \tag{3.22}$$

where τ_{h0} and τ_{e0} are lifetime parameters whose values depend on the type of trap and the volume density of trapping defects. n_1 and p_1 are parameters arising from the analysis which introduce a dependency of the recombination rate upon the energy of the trapping level, E_t:

$$n_1 = N_C \exp\left(\frac{E_t - E_c}{kT}\right) \tag{3.23}$$

$$n_1 p_1 = n_i^2 \tag{3.24}$$

Equation (3.23) is very similar in form to the expression for the electron concentration in terms of the Fermi-level energy as given by Eqs. (2.14) and (2.15). If τ_{e0} and τ_{h0} are of the same order of magnitude, it is not difficult to show that U will have its peak value when $n_1 \approx p_1$. This occurs if the defect level lies near the middle of the forbidden band gap. Therefore, impurities that introduce energy levels near midgap are very effective recombination centers.

3.4.5 Recombination at Surfaces

Surfaces represent rather severe defects in the crystal structure and are the site of many allowed states within the forbidden gap, as indicated in Fig. 3.12(b). Recombination can therefore occur very efficiently at surfaces by the mechanism described in 3.4.4. The net recombination rate per unit *area*, U_A, for a single-level surface state takes a form similar to Eq. (3.22):

$$U_A = \frac{S_{e0}S_{h0}(np - n_i^2)}{S_{e0}(n + n_1) + S_{h0}(p + p_1)} \tag{3.25}$$

where S_{e0} and S_{h0} are surface-recombination velocities. Again, surface-state levels lying near midgap are the most effective recombination centers.

3.5 BASIC EQUATIONS
OF SEMICONDUCTOR-DEVICE PHYSICS

3.5.1 Introduction

In previous sections, the relevant properties of semiconductors have been reviewed. This material will now be consolidated by synthesizing from it a set of basic equations capable of describing the operation of semiconductor devices. Solution of these equations allows the ideal characteristics of most semiconductor devices, including solar cells, to be determined. The equations will be written in one-dimensional form, with variations in the other two spatial dimensions neglected. Their three-dimensional form is similar except that spatial derivatives are replaced by the divergence operator for vector quantities (electric field, current density) and by the gradient operator for scalar quantities (concentrations, potentials).

3.5.2 Poisson's Equation

The first equation in the system may be familiar from electrostatics. It is Poisson's equation, one of Maxwell's equations (Ref. 3.5) which relates the divergence of the electric field to the space charge density, ρ. In one dimension, it takes the form

$$\frac{d\xi}{dx} = \frac{\rho}{\epsilon} \tag{3.26}$$

where ϵ is the material's permittivity. This equation is a differentiated form of Gauss's law, which may be more familiar.

Looking at the contributors to charge density in a semiconductor, electrons in the conduction band contribute a negative charge whereas holes give a positive charge. A donor impurity that is ionized (i.e., has its extra electron removed) has a net positive charge due to the unneutralized extra positive charge at the nucleus. Similarly, an ionized acceptor contributes a negative charge. Hence,

$$\rho = q(p - n + N_D^+ - N_A^-) \tag{3.27}$$

where p and n are the densities of holes and electrons, and N_D^+ and N_A^- are the densities of ionized donors and acceptors. Less specialized impurities and defects can also act as charge-storage centers, so corresponding terms should be included in Eq. (3.27). However, the volume density of such nonidealities is kept as small as possible in solar cell work, making the charge contributions relatively insignificant.

As mentioned in Chapter 2, most donors and acceptors are ionized under normal conditions, so that

$$N_D^+ \approx N_D$$
$$N_A^- \approx N_A$$

(3.28)

where N_D and N_A are the *total* density of donors and acceptors.

3.5.3 Current-Density Equations

In Chapter 2, it was seen that electrons and holes could contribute to current flow by drift and diffusion processes. Hence, the expressions for the total current densities of electrons and holes, J_e and J_h, become

$$J_e = q\,\mu_e\,n\,\xi + qD_e\,\frac{dn}{dx}$$

(3.29)

$$J_h = q\,\mu_h\,p\,\xi - qD_h\,\frac{dp}{dx}$$

The mobilities and diffusion constants are related through the Einstein relationships $[D_e = (kT/q)\mu_e; D_h = (kT/q)\mu_h]$.

3.5.4 Continuity Equations

The final equations in the system are "bookkeeping"-type equations which merely keep track of the number of electrons and holes in a system and ensure that none materialize "out of thin air."

Referring to the elemental volume of length δx and cross-sectional area A of Fig. 3.13, it can be stated that the net rate of increase of electrons in this volume is the rate at which they enter minus the rate at which they exit plus the rate at which they are generated in this volume minus the rate at which they recombine.

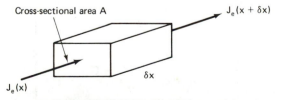

Figure 3.13.   Elemental volume for deriving the continuity equations for electrons.

But the rates of entering and exiting are proportional to the current densities at the respective faces of the elemental volume. Hence,

$$\text{rate of entering—rate exiting} = \frac{A}{q} \{ -J_e(x) - [-J_e(x + \delta x)] \}$$

$$= \frac{A}{q} \frac{dJ_e}{dx} \delta x \tag{3.30}$$

rate of generation—rate of recombination

$$= A \, \delta x (G - U) \tag{3.31}$$

where G is the net generation rate by external processes such as illumination by light, and U is the net recombination rate. Under steady-state conditions the net rate of increase must be zero, so that

$$\frac{1}{q} \frac{dJ_e}{dx} = U - G \tag{3.32}$$

Similarly, for holes

$$\frac{1}{q} \frac{dJ_h}{dx} = -(U - G) \tag{3.33}$$

3.5.5　Equation Set

The set of basic equations is

$$
\begin{array}{|c|}
\hline
\dfrac{d\xi}{dx} = \dfrac{q}{\epsilon} (p - n + N_D - N_A) \\[2ex]
J_e = q\mu_e n\xi + qD_e \dfrac{dn}{dx} \\[2ex]
J_h = q\mu_h p\xi - qD_h \dfrac{dp}{dx} \\[2ex]
\dfrac{1}{q} \dfrac{dJ_e}{dx} = U - G \\[2ex]
\dfrac{1}{q} \dfrac{dJ_h}{dx} = -(U - G) \\
\hline
\end{array} \tag{3.34}
$$

Subsidiary relations are also required for U and G. Expressions for these terms depend on the specific processes involved.

The Equations (3.34) form a coupled set of nonlinear differential equations for which it is not possible to find general analytical solutions. They can be solved numerically on a digital computer to give ideal properties for a range of possible semiconductor-device configurations. Examples of this technique applied to solar cells are contained in Refs. 3.6 to 3.8. It is also possible to get good solutions to these equations much more simply and with more insight into the physical principles involved by making a series of well-thought-out approximations. This technique is demonstrated in Chapter 4.

3.6 SUMMARY

Light made up of photons of energy larger than the forbidden band gap can be absorbed in a semiconductor by creating electron–hole pairs. In direct-band-gap semiconductor, the light is absorbed quickly. In indirect-band-gap materials, the emission or absorption of a phonon also is required for photon energies near that of the band gap. Hence, indirect-band-gap material absorbs weakly for such energies but becomes strongly absorbing for higher energies where direct transitions are also possible.

Recombination of carrier concentrations in excess of equilibrium values can occur through a variety of processes. Radiative recombination is the reverse of light absorption and is an important mechanism for direct-band-gap semiconductors. Auger recombination is important at high doping concentrations, whereas recombination through traps caused by impurities and defects is important for indirect-band-gap semiconductors and for those with an undeveloped supporting technology. These recombination processes occur in parallel. The total recombination rate is just the sum of the individual rates. The inverse of the net recombination lifetime is the sum of the inverses of the individual lifetimes. Recombination also occurs particularly effectively at semiconductor surfaces.

The end result of our review of the properties of semiconductors and the starting point for the analysis of the properties of solar cells is a system of coupled differential equations relating the spatial distributions of the quantities important in determining the internal operation of solar cells. Techniques for obtaining solutions to these equations are discussed in Chapter 4.

EXERCISES

3.1. Monochromatic light is normally incident on a flat silicon surface. Using the data of Fig. 3.1, calculate the fraction reflected at the following wavelengths: (a) 1000 nm; (b) 400 nm; (c) 300 nm. [*Note*: A relationship worth remembering between photon energy (hf) and wavelength in vacuum is λ (μm) = $1.24/hf$ (eV).]

3.2. (a) Monochromatic light of photon flux N photons per unit area per second is incident on a semiconductor surface and a fraction R is reflected. If the absorption coefficient of the semiconductor at this wavelength is α, what is an expression for the photon flux as a function of the distance, x, it progresses through the semiconductor?

(b) Given that each photon absorbed creates an electron–hole pair, calculate an expression in terms of the parameters above for the generation rate, G, of these pairs again as a function of distance through the semiconductor.

3.3. The peak photon flux in terrestrial sunlight arrives at wavelengths around 700 nm. Using the data of Figs. 3.4 and 3.6, compare the depth in Si and GaAs at which the photon flux at this wavelength has been reduced to 10% of its value immediately upon entering the respective semiconductors.

3.4. For a particular sample of semiconductor, the radiative recombination lifetime for minority carriers is calculated as 100 μs, the Auger recombination lifetime as 50 μs, and the lifetime for trapping processes as 10 μs. Assuming that no other process is available for recombination, what is the net lifetime in this material?

3.5. A sample of n-type silicon is illuminated by light so that the electron density is constant at 10^{22} per m^3 and the hole density is also constant at 10^{15} per m^3. Determine the effectiveness of recombination by traps as a function of trap energy by calculating the recombination rate of electrons and holes for the case where the trap is assumed located at each of the following energies below the conduction-band edge: (a) 0.03 eV; (b) 0.3 eV; (c) 0.5 eV; (d) 0.8 eV; (e) 1.0 eV. Assume the density and capture cross sections of the trap in each case is such that τ_{e0} and τ_{h0} both equal 1 μs. Use the values N_C = 3 $\times$ 10^{25} m^{-3}, n_i = 1.5 $\times$ 10^{16} m^{-3}, and kT/q = 26 mV.

3.6. In terms of the electronic properties of semiconductors, explain why the absorption coefficient increases with increasing photon energy for energies near the semiconductor band gap.

REFERENCES

[3.1]   O. S. HEAVENS, *Optical Properties of Thin Solid Films* (London: Butterworths, 1955).

[3.2]   J. I. PANKOVE, *Optical Processes in Semiconductors* (Englewood Cliffs, N.J.: Prentice-Hall, 1971).

[3.3]   K. RAJKANAN, R. SINGH, AND J. SHEWCHUN, "Absorption Coefficient of Silicon for Solar Cell Calculations," *Solid-State Electronics 22* (1979), 793.

[3.4] C. T. SAH, R. N. NOYCE, AND W. SHOCKLEY, "Carrier Generation and Recombination in *p-n* Junctions and *p-n* Junction Characteristics," *Proceedings of the IRE 45* (1957), 1228.

[3.5] S. M. SZE, *Physics of Semiconductor Devices* (New York: Wiley, 1969).

[3.6] P. M. DUNBAR AND J. R. HAUSER, "A Study of Efficiency in Low Resistivity Silicon Solar Cells," *Solid-State Electronics 19* (1976), 95–102.

[3.7] J. G. FOSSUM, "Computer-Aided Numerical-Analysis of Silicon Solar Cells," *Solid-State Electronics 19* (1976), 269–277.

[3.8] M. A. GREEN, F. D. KING, AND J. SHEWCHUN, "Minority Carrier MIS Tunnel Diodes and Their Application to Electron- and Photo-voltaic Energy Conversion," *Solid-State Electronics 17* (1974), 551–561.

Chapter

p-n JUNCTION DIODES

4.1 INTRODUCTION

Regions of semiconductors doped with donor impurities have an increased number of electrons in the conduction band at normal temperatures and are known as *n-type material*. Those doped with acceptors are known as *p-type*. The most common solar cells are essentially just very large area *p-n junction diodes*, where such a diode is formed by making a junction between *n*-type and *p*-type regions. In the present chapter, the basic properties of such a junction will be analyzed both when in the dark and when illuminated.

The basic device requirement for photovoltaic energy conversion is an electronic asymmetry in the semiconductor structure. Figure 4.1(a) demonstrates that a *p-n* junction has the required asymmetry. *n*-Type regions have large electron densities but small hole densities. Hence, electrons flow readily through such material but holes find it very difficult. Exactly the opposite is true for *p*-type material. When illuminated, excess electron–hole pairs are generated by light throughout the material. The inherent asymmetry in the

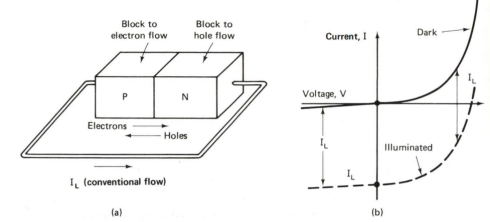

Block to electron flow   Block to hole flow

P   N

Electrons ——→
←—— Holes

I_L (conventional flow)

(a)

Current, I   Dark ——→

Voltage, V

I_L

I_L

I_L

Illuminated

(b)

Figure 4.1. (a) Asymmetrical properties of a p-n junction diode. These cause a net current flow in an external lead connecting the p-type and n-type regions when the junction is illuminated. (b) This light-generated current is superimposed upon the normal rectifying current–voltage characteristics of the diode. This results in a region in the fourth quadrant where electrical power can be extracted from the device.

carrier transport properties encourages a flow of generated electrons from the p-type region to the n-type, and a flow of holes in the opposite direction. When the illuminated p-n junction is electrically shorted, a current will flow in the short-circuiting lead. It will be shown in this chapter that this light-generated current is superimposed upon the normal diode-rectifying characteristics to give an operating region where power can be extracted from the cell as indicated in Fig. 4.1(b).

4.2 ELECTROSTATICS OF p-n JUNCTIONS

Consider isolated pieces of n-type and p-type semiconducting material as shown in Fig. 4.2. If these are brought together in a conceptual experiment, it would be expected that electrons would flow from regions of high concentration (n-type side) to regions of low concentration (p-type side) and similarly for holes. However, electrons leaving the n-type side will create a charge imbalance in this side by exposing ionized donors (positive charge). Similarly, holes leaving the p-type side will expose negative charge. These exposed charges

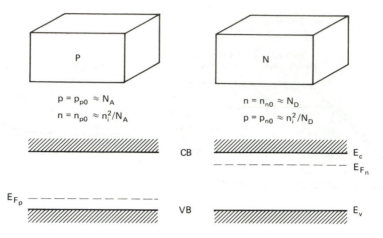

$p = p_{p0} \approx N_A$

$n = n_{p0} \approx n_i^2/N_A$

$n = n_{n0} \approx N_D$

$p = p_{n0} \approx n_i^2/N_D$

CB

VB

E_{F_p}

E_c

E_{F_n}

E_v

Figure 4.2. Isolated pieces of *p*-type and *n*-type semiconductor material with corresponding energy-band diagrams.

will set up an electric field that will oppose the natural diffusion tendency of the electrons and holes and an equilibrium situation will be obtained.

The characteristics of the equilibrium situation can be found by considering Fermi levels. *A system in thermal equilibrium can have only one Fermi level.*

Far enough away from the metallurgical junction, conditions could be expected to remain unperturbed from those in isolated material. Referring to Fig. 4.3, this means there must be a transition region near the junction in which a potential change, ψ_0, occurs. The value of ψ_0 can be found from this figure:

$$q\psi_0 = E_g - E_1 - E_2 \tag{4.1}$$

The expressions for E_1 and E_2 were derived in Eqs. (2.28) and (2.30) and are also indicated in Fig. 4.3. Hence,

$$q\psi_0 = E_g - kT \ln\left(\frac{N_V}{N_A}\right) - kT \ln\left(\frac{N_C}{N_D}\right)$$

$$= E_g - kT \ln\left(\frac{N_C N_V}{N_A N_D}\right) \tag{4.2}$$

but, from Eq. (2.17),

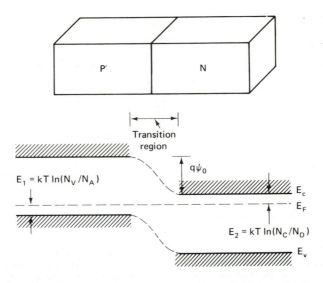

Figure 4.3. A *p-n* junction formed by bringing the isolated *p*-type and *n*-type regions together. Also shown is the corresponding energy-band diagram at thermal equilibrium.

$$n_i^2 = N_C N_V \, \exp\left(-\frac{E_g}{kT}\right)$$

Therefore,

$$\psi_0 = \frac{kT}{q} \ln\left(\frac{N_A N_D}{n_i^2}\right) \tag{4.3}$$

An applied voltage, V_a, will change the potential difference between the two sides of the diodes by V_a. Hence, the potential across the transition region will become ($\psi_0 - V_a$).

It is instructive to plot the carrier concentrations corresponding to Fig. 4.3. These concentrations depend on the exponential of the energy difference between the Fermi level and the respective band. The resulting plot is shown in Fig. 4.4 on a logarithmic scale. A corresponding plot of the space-charge density, ρ, as given by Eq. (3.27), is shown as the dashed line of Fig. 4.5(a). The rapid change of ρ near the edge of the depletion region leads to *approximation 1*, the *depletion approximation*.

In this approximation, the device is divided into two types of regions: *quasi-neutral regions* where the space-charge density is as-

65

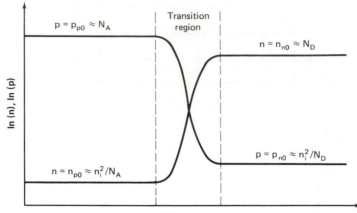

Figure 4.4. Plot of the natural logarithms of the electron and hole concentrations corresponding to Fig. 4.3. Since these concentrations depend exponentially on the energy between the Fermi level and the respective band, the shapes of the distributions on a logarithmic scale are linearly related to that of Fig. 4.3.

sumed zero throughout and a *depletion region* where the carrier concentrations are assumed so small that the only contribution to space-charge density comes from the ionized dopants. This approximation essentially just sharpens up the space-charge distribution, as indicated by the solid line in Fig. 4.5(a).

With this approximation, it is a simple matter to find the electric field and potential distribution across the depletion region, as shown in Fig. 4.5(b) and (c). The space-charge distribution is just integrated successively, remembering that electric field strength is the negative gradient of potential. The results for the maximum field strength in the depletion region, ξ_{max}, the width of the depletion region, W, and the distance this region extends on either side of the junction, l_n and l_p, are (Ref. 4.1)

$$\xi_{max} = -\left[\frac{2q}{\epsilon}(\psi_0 - V_a)\Bigg/\left(\frac{1}{N_A} + \frac{1}{N_D}\right)\right]^{1/2}$$

$$W = l_n + l_p = \left[\frac{2\epsilon}{q}(\psi_0 - V_a)\left(\frac{1}{N_A} + \frac{1}{N_D}\right)\right]^{1/2} \tag{4.4}$$

$$l_p = W\frac{N_D}{N_A + N_D} \qquad l_n = W\frac{N_A}{N_A + N_D}$$

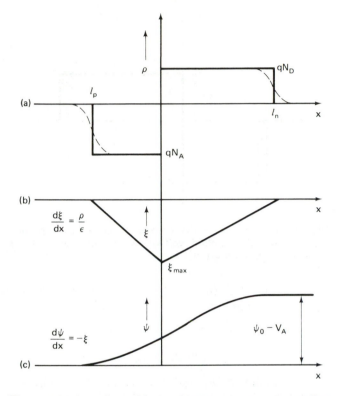

Figure 4.5. (a) Space-charge density corresponding to Fig. 4.4. The dashed line shows the actual distribution while the solid line shows the assumed distribution in the depletion approximation. (b) Corresponding electrical field strength. (c) Corresponding potential distribution.

4.3 JUNCTION CAPACITANCE

It is a very easy matter to detect the presence of a depletion region in a *p-n* diode and to measure its width. In the depletion approximation, a change in applied voltage will cause a change in stored charge right at the edges of the region, as indicated in Fig. 4.6. This is identical to the situation of a parallel-plate capacitor of plate separation, W. Hence, the depletion-region capacitance, C, is

$$C = \frac{\epsilon A}{W} \qquad (4.5)$$

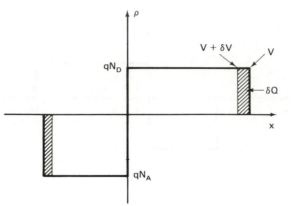

Figure 4.6. Change in the charge stored in the depletion region when the applied voltage is incrementally altered (depletion approximation).

where W is given by Eq. (4.4). If one side of the diode is heavily doped, Eq. (4.5) reduces to:

$$\frac{C}{A} = \left[\frac{q \epsilon N}{2(\psi_0 - V_a)} \right]^{1/2} \tag{4.6}$$

where N is the smaller of N_A and N_D. Under reverse bias, depletion-region capacitance dominates the total diode capacitance. Hence, measuring C as a function of reverse bias to a diode or solar cell and plotting $1/C^2$ versus V_a will allow N, the doping density on the lightly doped side of the diode, to be found. A similar technique can also be used (Ref. 4.2) to calculate the spatial variation of the dopant density in cases where this quantity is not constant.

4.4 CARRIER INJECTION

The next calculation involves finding the concentration of carriers at the edge of the depletion region as a function of bias. Referring to Fig. 4.7, values are sought for the concentrations $n_{p\,a}$ and $p_{n\,b}$.

At zero bias, their values are already known (Fig. 4.4):

$$p_{n\,b} = p_{n\,0} = p_{p\,0} \exp\left(-\frac{q\psi_0}{kT} \right) \approx \frac{n_i^2}{N_D}$$

$$n_{p\,a} = n_{p\,0} = n_{n\,0} \exp\left(-\frac{q\psi_0}{kT} \right) \approx \frac{n_i^2}{N_A}$$

$$\tag{4.7}$$

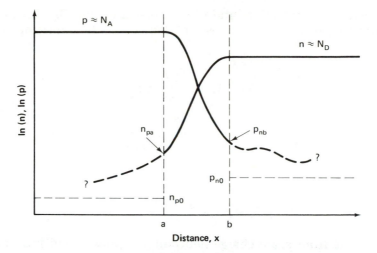

Figure 4.7. Plot of carrier concentrations when a voltage is applied to the *p-n* junction. In the text, expressions are found for the minority carrier concentrations n_{pa} and p_{nb} at the edge of the junction depletion region. Subsequently, the precise form of the distributions shown dashed are also calculated.

Within the depletion region, there are present both the highest electric field strengths and concentration gradients. The net current flow through such regions is actually the small difference between two large terms. For holes,

$$J_h = q\mu_h p\xi - qD_h \frac{dp}{dx} \tag{4.8}$$

Both the drift and diffusion terms are large but opposing. At zero bias they balance. At moderate bias points, the net current flow is a small difference between these two much larger terms. This leads to *approximation 2*—that, within depletion regions,

$$q\mu_h p\xi \approx qD_h \frac{dp}{dx} \tag{4.9}$$

In other words,

$$\xi \approx \frac{kT}{q} \frac{1}{p} \frac{dp}{dx} \tag{4.10}$$

69

making use of Einstein's relationship between μ_h and D_h. Integrating
the negative of both sides of Eq. (4.10) across the depletion region
gives

$$\psi_0 - V_a = -\frac{kT}{q}\ln p\,\Big|_a^b$$

$$= \frac{kT}{q}\ln\frac{p_{pa}}{p_{nb}} \tag{4.11}$$

or, rearranging,

$$p_{nb} = p_{pa}e^{-q\psi_0/kT}e^{qV_a/kT} \tag{4.12}$$

But from space-charge neutrality at point a and introducing
approximation 3, that only cases where the carriers in the minority
have a much lower concentration than those in the majority will be
considered ($p_{pa} \gg n_{pa}$, $n_{na} \gg p_{na}$), gives

$$p_{pa} = N_A + n_{pa} \quad (\text{where } n_{pa} \text{ is small})$$

$$\approx p_{p0} \approx p_{n0}e^{q\psi_0/kT} \tag{4.13}$$

Hence,

$$\boxed{\begin{aligned} p_{nb} &= p_{n0}e^{qV_a/kT} = \frac{n_i^2}{N_D}e^{qV_a/kT} \\[2mm] n_{pa} &= n_{p0}e^{qV_a/kT} = \frac{n_i^2}{N_A}e^{qV_a/kT} \end{aligned}} \tag{4.14}$$

Therefore, the concentration of carriers in the minority
(*minority carriers*) at the edge of the depletion region increases exponentially with applied voltage. The process by which this concentration is controlled by the bias across the junction is known as
minority-carrier injection.

4.5   DIFFUSIVE FLOW IN QUASI-NEUTRAL REGIONS

Carriers can flow by drift and diffusion. If a uniformly doped region
of semiconductor material is quasi-neutral (space-charge density ap-

proximately zero) and minority-carrier flows are not insignificant, minority carriers will flow predominantly by diffusion.

Proof (reductio ad absurdum): Consider n-type quasi-neutral material where $n \gg p$ and the case where minority-carrier flows are not insignificant (i.e., the case where $J_e \not\gg J_h$, with the symbol $\not\gg$ is used to mean "not a lot greater than")
We have

$$J_e = q\mu_e n\xi + qD_e \frac{dn}{dx}$$

$$J_h = q\mu_h p\xi - qD_h \frac{dp}{dx}$$

$$(4.15)$$

$$p - n + N_D \approx 0 \qquad \text{(quasi-neutral)}$$

Differentiating the latter equation, remembering that N_D is constant, yields

$$\frac{dp}{dx} \approx \frac{dn}{dx} \tag{4.16}$$

Assume that the drift component of minority-carrier flow is *not* negligible, that is,

$$|q\mu_h p\xi| \not\ll \left|qD_h \frac{dp}{dx}\right| \tag{4.17}$$

Since $n \gg p$, it follows, applying Eqs. (4.17) and then (4.16), that

$$|q\mu_h n\xi| \gg \left|qD_h \frac{dp}{dx}\right| \gg \left|qD_h \frac{dn}{dx}\right|$$

Equivalently,

$$|q\mu_e n\xi| \gg \left|qD_e \frac{dn}{dx}\right| \tag{4.18}$$

Also, since μ_e and μ_h are of comparable magnitude,

$$|q\mu_e n\xi| \gg |q\mu_h p\xi| \tag{4.19}$$

Combining Eqs. (4.17) to (4.19) leads to the conclusion:

$$J_e \gg J_h$$

This violates one of the original conditions. Therefore, the assumption of Eq. (4.17) is incorrect. This leads to the result

$$\left| q\,\mu_h\,p\,\xi \right| \ll \left| qD_h\,\frac{dp}{dx} \right|$$

That is, *minority carriers* in *quasi-neutral regions* flow predominantly by diffusion, subject to the conditions previously mentioned. Hence, *approximation 4* is

$$J_h = -qD_h\,\frac{dp}{dx} \qquad (n\text{-type quasi-neutral region})$$

$$\tag{4.20}$$

$$J_e = qD_e\,\frac{dn}{dx} \qquad (p\text{-type quasi-neutral region})$$

Basically, the small number of minority carriers compared to majority carriers shields them from the effect of an electric field. In the following sections, the relevance of this in connection with finding current flows in *p-n* diodes will become apparent.

4.6 DARK CHARACTERISTICS

4.6.1 Minority Carriers in Quasi-Neutral Regions

To summarize progress to date, it has been shown that a reasonable approximation for analysis is to divide the diode up into depletion and quasi-space-charge neutral regions. It has been found that the minority-carrier concentration at the edge of the depletion region depends exponentially on the voltage applied to the diode. This information is summarized in Fig. 4.7. Moreover, it has been shown that, when quasi-neutral regions are uniformly doped and majority carrier currents small, minority carriers flow primarily by diffusion. This will allow the distributions shown dashed in Fig. 4.7 to be calculated.

On the n-type side of the diode,

$$J_h = -qD_h \frac{dp}{dx} \tag{4.21}$$

while the continuity equation gives

$$\frac{1}{q}\frac{dJ_h}{dx} = -(U - G) \tag{4.22}$$

Explicit expressions for the recombination rate were given in Chapter 3 for several recombination mechanisms. From the definition of carrier lifetimes of Eq. (3.19), the recombination rate in the n-type region can be put in the form

$$U = \frac{\Delta p}{\tau_h} \tag{4.23}$$

where Δp is the excess concentration of minority-carrier holes equal to the total concentration, p_n, minus the equilibrium concentration, p_{n0}. τ_h is the minority-carrier lifetime which can be regarded as a constant, at least for small disturbances from equilibrium. Combining the three equations above gives

$$D_h \frac{d^2 p_n}{dx^2} = \frac{p_n - p_{n0}}{\tau_h} - G \tag{4.24}$$

In the dark, $G = 0$. Also, $d^2 p_{n0}/dx^2 = 0$. Hence, Eq. (4.24) simplifies to

$$\frac{d^2 \Delta p}{dx^2} = \frac{\Delta p}{L_h^2} \tag{4.25}$$

where

$$L_h = \sqrt{D_h \tau_h} \tag{4.26}$$

L_h has the dimensions of length and is known as the *diffusion length*. It will become apparent that this is a very important parameter in solar cell work. The general solution to Eq. (4.25) is

$$\Delta p = A e^{x/L_h} + B e^{-x/L_h} \tag{4.27}$$

The constants A and B can be found by applying two boundary conditions as follows:

1. At $x = 0$, $p_{nb} = p_{n0} e^{q V/kT}$.
2. p_n finite as $x \to \infty$. Therefore, $A = 0$.

These boundary conditions give the particular solution:

$$p_n(x) = p_{n0} + p_{n0}[e^{q V/kT} - 1] e^{-x/L_h} \qquad (4.28)$$

Similarly,

$$n_p(x') = n_{p0} + n_{p0}[e^{q V/kT} - 1] e^{-x'/L_e} \qquad (4.29)$$

where x' is defined in Fig. 4.8(b).

These solutions for minority-carrier concentrations throughout the device are plotted in Fig. 4.8(a). In the quasi-neutral regions, the majority-carrier concentrations must have a corresponding change in their distributions to maintain space-charge neutrality, also as shown in Fig. 4.8(a). Even though the absolute changes are the same, the relative changes in the majority changes are very much smaller, as indicated in the logarithmic plot of Fig. 4.8(b).

4.6.2 Minority-Carrier Currents

It is a very simple matter to calculate minority-carrier current flows once the carrier distributions are known. Since currents are

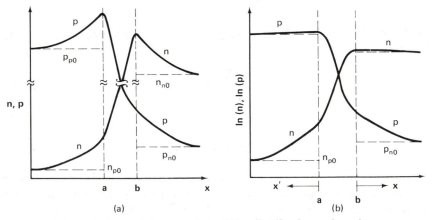

Figure 4.8. (a) Linear plot of the distributions of carriers throughout the p-n junction diode under forward bias. (b) Corresponding semilogarithmic plot. Note the differences with respect to majority carriers.

diffusive in quasi-neutral regions, on the n-type side:

$$J_h = -qD_h \frac{dp}{dx} \tag{4.30}$$

Substituting Eq. (4.28) gives

$$J_h(x) = \frac{qD_h p_{n0}}{L_h} (e^{qV/kT} - 1)e^{-x/L_h} \tag{4.31}$$

Similarly, in the p-type region,

$$J_e(x') = \frac{qD_e n_{p0}}{L_e} (e^{qV/kT} - 1)e^{-x'/L_e} \tag{4.32}$$

The current distributions resulting from these expressions are plotted in Fig. 4.9(a). In order to calculate the total current flow in the diode, it is necessary to know both the electron and hole components at the same point. Considering current flows in the depletion region, the continuity equations give

$$\frac{1}{q}\frac{dJ_e}{dx} = U - G = -\frac{1}{q}\frac{dJ_h}{dx} \tag{4.33}$$

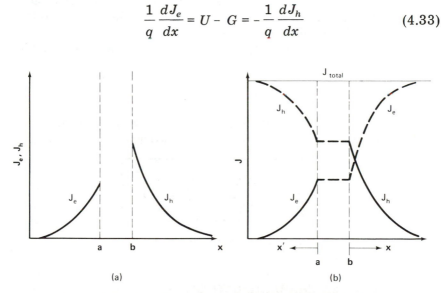

(a)   (b)

Figure 4.9. (a) Minority-carrier current densities in a p-n junction diode corresponding to Fig. 4.8. (b) Distribution of minority, majority, and total current densities in the diode, neglecting recombination in the depletion region.

Hence, the magnitude of the change in current across the depletion region is

$$\delta J_e = |\delta J_h| = q \int_{-W}^{0} (U - G)\, dx \qquad (4.34)$$

W is generally much less than L_e and L_h, the characteristic decay lengths of J_e and J_h. This indicates that Fig. 4.9(a) is grossly out of proportion. Since W is small, a reasonable approximation is to assume that the integral involved in Eq. (4.34) is negligible, so that $\delta J_e = |\delta J_h| \approx 0$. It follows that J_e and J_h are essentially constant across the depletion region as shown in Fig. 4.9(b). This approximation, *approximation 5*, would appear more reasonable if W were drawn to scale. The total current can now be found since both J_e and J_h are known at all points in the depletion region. Hence,

$$
\begin{aligned}
J_{\text{total}} &= J_e \big|_{x'=0} + J_h \big|_{x=0} \\
&= \left(\frac{qD_e n_{p0}}{L_e} + \frac{qD_h p_{n0}}{L_h} \right) (e^{qV/kT} - 1)
\end{aligned}
\qquad (4.35)
$$

Since J_{total} is constant with position, it is now possible to complete the distributions of J_e and J_h throughout the diode as shown by the dashed lines of Fig. 4.9(b).

The result of the analysis has been the derivation of the ideal diode law:

$$I = I_0 (e^{qV/kT} - 1) \qquad (4.36)$$

The important point as far as this book is concerned is the expression derived for the saturation current density:

$$\boxed{I_0 = A \left(\frac{qD_e n_i^2}{L_e N_A} + \frac{qD_h n_i^2}{L_h N_D} \right)} \qquad (4.37)$$

where A is the cross-sectional area of the diode.

4.7 ILLUMINATED CHARACTERISTICS

Next the characteristics of a *p-n* diode when illuminated will be explored. For mathematical simplicity, an idealized case will be assumed where the generation rate of electron–hole pairs by the illumination will be assumed uniform throughout the device. This would correspond closely to a specific physical situation where the cell was illuminated by long-wavelength light consisting of photons of energy close to that of the semiconductor band gap. Such light would be only weakly absorbed and the volume generation rate of electron-hole pairs would be approximately constant over the distances characteristically involved. It is emphasized that this case of uniform generation rate *does not correspond to the situation in actual solar energy conversion.* More realistic situations will be treated using a different approach in subsequent chapters.

Problem: Derive the ideal current-voltage characteristics of a p-n junction diode when illuminated by light such that the volume rate of generation of electron-hole pairs by the light, G, is constant throughout the device.

The analysis closely parallels that of the diode in the dark. To consolidate this material, the reader is encouraged to attempt to work through the problem without reference to the worked solution that follows.

Worked Solution: The reader should first be satisfied that approximations 1 through 4 and the intermediate results leading from them are equally valid regardless of whether the device is illuminated or not. This being the case, Eq. (4.24) is still valid, but in this case G is not zero but a constant. Hence, on the n-type side:

$$\frac{d^2 \Delta p}{dx^2} = \frac{\Delta p}{L_h^2} - \frac{G}{D_h} \tag{4.38}$$

Since G/D_h is constant, the corresponding general solution is

$$\Delta p = G\tau_h + Ce^{x/L_h} + De^{-x/L_h} \tag{4.39}$$

The boundary conditions remain unchanged from the analysis of the diode in the dark. This gives the particular solution

$$p_n(x) = p_{n0} + G\tau_h + [p_{n0}(e^{qV/kT} - 1) - G\tau_h]\, e^{-x/L_h} \tag{4.40}$$

with a similar expression for $n_p(x')$ as plotted in Fig. 4.10.

The corresponding current density is

$$J_h(x) = \frac{qD_h p_{n0}}{L_h}(e^{qV/kT} - 1)e^{-x/L_h} - qGL_h e^{-x/L_h} \tag{4.41}$$

with a similar expression for $J_e(x')$.

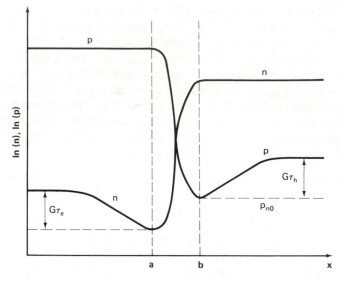

Figure 4.10. Distribution of carriers through a *p-n* junction when short-circuited under infrared illumination (generation rate assumed uniform through diode).

Again neglecting the effect of recombination in the depletion region (approximation 5) but including the effect of generation in this region in this case gives the change in current density across this region as

$$|\delta J_e| = |\delta J_h| = qGW \tag{4.42}$$

Hence, proceeding as before gives the following result for the current–voltage characteristics:

$$I = I_0 \left(e^{qV/kT} - 1\right) - I_L \tag{4.43}$$

where I_0 retains the value of Eq. (4.37) and I_L has the value

$$I_L = qAG(L_e + W + L_h) \tag{4.44}$$

This result is plotted in Fig. 4.11. Note that the illuminated characteristics are merely the dark characteristics shifted down by a current I_L. This gives a region in the fourth quadrant of this plot where power can be extracted from the diode.

Note that the form of Eq. (4.44) suggests a conclusion that will be confirmed later. The light-generated current, I_L, has a value

78

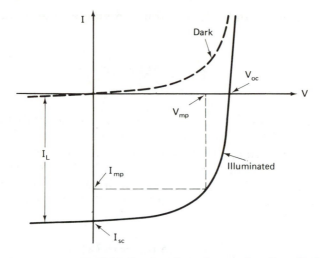

Figure 4.11. Terminal properties of a *p-n* junction diode in the dark and when illuminated.

equal to that expected if all the carriers generated by light in the depletion region of the diode and within a minority carrier diffusion length on either side were to contribute to it. The depletion region and the generally much larger volume of material lying within a diffusion length of either side of it is indeed the "active" collection region of a *p-n* junction solar cell.

4.8 SOLAR CELL OUTPUT PARAMETERS

Three parameters are usually used to characterize solar cell outputs (Fig. 4.11).

One of these is the *short-circuit current*, I_{sc}. Ideally, this is equal to the light-generated current I_L. A second parameter is the *open-circuit voltage*, V_{oc}. Setting I to zero in Eq. (4.43) gives the ideal value:

$$V_{oc} = \frac{kT}{q} \ln \left(\frac{I_L}{I_0} + 1 \right) \qquad (4.45)$$

V_{oc} is determined by the properties of the semiconductor by virtue of its dependence on I_0. The power output for any operating point in the fourth quadrant is equal to the area of the rectangle indicated in Fig. 4.11. One particular operating point (V_{mp}, I_{mp}) will

79

maximize this power output. The third parameter, the fill factor, FF, is defined as

$$FF = \frac{V_{mp}I_{mp}}{V_{oc}I_{sc}} \qquad (4.46)$$

It is a measure of how "square" the output characteristics are. For cells of reasonably efficiency, it has a value in the range 0.7 to 0.85. Ideally, it is a *function only of the open-circuit voltage*, V_{oc}. Defining a normalized voltage, v_{oc}, as $V_{oc}/(kT/q)$, the ideal (maximum) value of FF is shown in Fig. 4.12. An empirical expression describing this relationship to about four significant digits for $v_{oc} > 10$ is (see also Section 5.4.4)

$$FF = \frac{v_{oc} - \ln(v_{oc} + 0.72)}{v_{oc} + 1} \qquad (4.47)$$

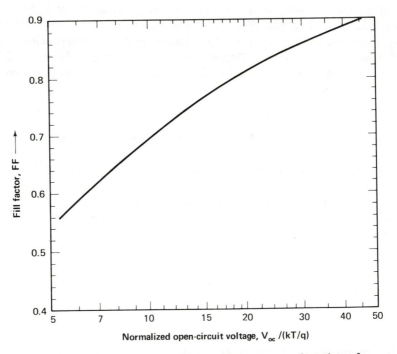

Figure 4.12. Ideal value of the fill factor as a function of the open-circuit voltage normalized to the thermal voltage, kT/q.

The energy-conversion efficiency, η, is then given by

$$\eta = \frac{V_{mp}I_{mp}}{P_{in}} = \frac{V_{oc}I_{sc}FF}{P_{in}} \tag{4.48}$$

where P_{in} is the total power in the light incident on the cell. Energy-conversion efficiencies of commercial solar cells generally lie in the 12 to 14% range.

4.9 EFFECT OF FINITE CELL DIMENSIONS ON I_0

The diode saturation current, I_0, determines V_{oc} as indicated by Eq. (4.45). In deriving Eq. (4.37) for I_0, it was implicitly assumed that the diode extended an unlimited distance on either side of the junction. Actual devices will not. A solar cell of finite dimensions is indicated in Fig. 4.13.

This modifies the value of the saturation current, I_0. The modified value will depend on the surface-recombination velocities (Section 3.4.5) at the exposed surfaces. Two limiting cases are of interest: (1) when this velocity is very high, approaching infinity; and (2) when it is very low, approaching zero. In the former case the excess minority carrier concentration is zero at the surface. In the latter, the minority carrier current flow into the surface is zero. Applying these as boundary conditions allows the modified expression for I_0 to be found (Exercise 4.3). It takes the form[1]

$$I_0 = A\left(\frac{qD_e n_i^2}{L_e N_A} * F_P + \frac{qD_h n_i^2}{L_h N_D} * F_N\right) \tag{4.49}$$

If the surface on the p-type side of the device is a high recombination velocity surface, F_P has the form

$$F_P = \coth\left(\frac{W_P}{L_e}\right) \tag{4.50}$$

---

[1]General expressions for F_N and F_P are (Ref. 4.3)

$$F_N = \frac{S_h \cosh(W_N/L_h) + D_h/L_h \sinh(W_N/L_h)}{D_h/L_h \cosh(W_N/L_h) + S_h \sinh(W_N/L_h)}$$

$$F_P = \frac{S_e \cosh(W_P/L_e) + D_e/L_e \sinh(W_P/L_e)}{D_e/L_e \cosh(W_P/L_e) + S_e \sinh(W_P/L_e)}$$

where S_e and S_h are the surface-recombination velocities of the respective surfaces as introduced in Section 3.4.5.

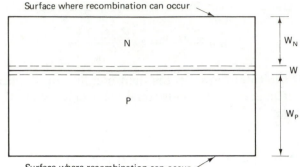

Figure 4.13. Basic solar cell, defining important dimensions.

where W_P is defined in Fig. 4.13. A corresponding expression will hold for F_N if the corresponding surface is also a high-recombination-velocity surface. If it happens to be a *low-recombination-velocity* surface, then F_N would be given by

$$F_N = \tanh\left(\frac{W_N}{L_h}\right) \qquad (4.51)$$

A similar expression would hold for F_P if the *p*-type region were also to have a low-recombination-velocity surface.

Note that the smallest value of I_0 and hence the largest V_{oc} will occur if both surfaces possess low recombination velocities.

4.10   SUMMARY

By making a series of approximations, the general system of equations describing solar cell operation can be reduced to a tractable form. This approach has allowed the ideal form of the dark and illuminated characteristics of solar cells to be found.

Under illumination, the dark current–voltage curves ideally are displaced downward by the light-generated current. The active region of a solar cell for collecting the latter current is the junction depletion region and the volume of cell material within a minority carrier diffusion length of this region.

The parameters used to characterize the terminal properties of a solar cell are the short-circuit, I_{sc}, the open-circuit voltage, V_{oc}, and the fill factor, FF. The conditions at the surfaces of a solar cell, necessarily of finite extent, influence the open-circuit voltage and will be shown in subsequent chapters also to affect the short-circuit current.

EXERCISES

4.1. A. *p-n* junction diode is uniformly doped with 10^{24} dopants/m^3 on the *p*-type side and 10^{22} dopants on the *n*-type side. At 300 K, calculate the maximum electric field strength, the width of the depletion region, and the junction capacitance per unit area for the following bias conditions: (a) zero bias; (b) 0.4 V forward bias; (c) 10 V reverse bias.

4.2. The derivation of Section 4.4 assumes that in quasi-neutral regions, carriers in the minority have a much lower concentration than majority carriers. Find an expression for the maximum applied voltage at which this can be valid.

4.3. (a) Consider the case of a cell of finite dimensions as shown in Fig. 4.13. Derive expressions for the electron concentration on the *p*-type side in the dark as a function of applied voltage for the two extreme cases where the back surface has a very high recombination velocity and a very low recombination velocity. Sketch and compare these distributions when the width of the *p*-type region is very much less than a minority-carrier diffusion length.

 (b) Referring to this sketch, indicate which distribution would give the smallest contribution to the diode saturation current and hence the larger open-circuit voltage under illumination if other parameters of the diode are identical.

4.4. The following is a further example of the analysis technique of this chapter, which is not as demanding mathematically. Consider a cell of finite dimensions which are much smaller than corresponding minority-carrier diffusion lengths. The recombination velocities along the front and rear surface of this cell are given as very high, assumed infinite. In this case, it will be a good approximation to neglect bulk recombination compared to surface recombination (i.e., the recombination rate U can be assumed zero throughout bulk regions). Find an expression for the saturation current density of this diode and for the short-circuit current for the *special case* where the generation rate of electron–hole pairs by light, G, is constant throughout the cell.

4.5. When the cell temperature is 300 K, a certain silicon cell of 100-cm^2 area gives an open-circuit voltage of 600 mV and a short-circuit current output of 3.3 A under 1-kW/m^2 illumination. Assuming that the cell behaves ideally, what is its energy-conversion efficiency at the maximum power point?

REFERENCES

[4.1] A. S. GROVE, *Physics and Technology of Semiconductor Devices* (New York: Wiley, 1967), p. 158.

[4.2] Ibid., pp. 169–172.

[4.3] J. P. McKELVEY, *Solid State and Semiconductor Physics* (New York: Harper & Row, 1966), p. 422.

Chapter

5

EFFICIENCY LIMITS, LOSSES, AND MEASUREMENT

5.1 INTRODUCTION

Sunlight incident on a solar cell creates electron–hole pairs within the semiconducting material making up the cell. The cell has an asymmetrical electronic structure which causes the electrons and holes making up these generated electron–hole pairs to separate and create current flow in a load connected between the cell terminals. In the present chapter, energy-conversion efficiency limits are discussed for this process as well as the effects of various nonidealities on efficiency. Techniques for measuring the efficiency of photovoltaic devices are also described.

5.2 EFFICIENCY LIMITS

5.2.1 General

In Chapter 4, it was seen that three parameters could be used to characterize the performance of a *p-n* junction solar cell. These are the *open-circuit voltage* (V_{oc}), the *short-circuit current* (I_{sc}), and the *fill factor* (FF) (Fig. 4.11). It was also indicated that the maximum value of the fill factor is a function of V_{oc}. Hence, only the ideal limits of I_{sc} and V_{oc} need to be examined.

5.2.2   Short-Circuit Current

It is relatively easy to calculate the upper limit to the short-circuit current obtainable from any selected solar cell material. Under ideal conditions, each photon incident on the cell of energy greater than the band gap gives rise to one electron flowing in the external circuit.[1] Hence, to calculate the maximum I_{sc}, the photon flux in sunlight must be known. This can be calculated from the energy distribution of sunlight (Chapter 1) by dividing the energy content at a given wavelength by the energy of an individual photon (hf or hc/λ) of this wavelength. The results of such a calculation are shown in Fig. 5.1(a) for AM0 radiation and the standard terrestrial AM1.5 radiation mentioned in Chapter 1.

The maximum I_{sc} is then found by integrating these distributions from low wavelengths up to the maximum wavelength for which electron–hole pairs can be generated for a given semiconductor. [A relationship between photon energy in eV and its wavelength in micrometers worth remembering is E (eV) = $1.24/\lambda$ (μm). For silicon, the band gap is about 1.1 eV, so that λ corresponding to this is 1.13 μm.] The resulting upper limits on short-circuit current density are shown in Fig. 5.1(b).

It is not surprising that as the band gap decreases, the short-circuit current density increases. More photons have the energy required to create an electron–hole pair as the band gap becomes smaller.

5.2.3   Open-Circuit Voltage and Efficiency

The fundamental limitations on the open-circuit voltages of solar cells are not as clearly defined. In Chapter 4, it was shown that, for an ideal *p-n* junction cell, V_{oc} was given by

$$V_{oc} = \frac{kT}{q} \ln \left(\frac{I_L}{I_0} + 1 \right) \tag{5.1}$$

---

[1] Very high energy photons of energy several times the band-gap energy can create an electron–hole pair where the electron has sufficient energy above the conduction-band edge to create a second electron–hole pair by *impact ionization* (Section 3.4.3). There are not many such photons in sunlight, so this mechanism is not of significance in solar cells.

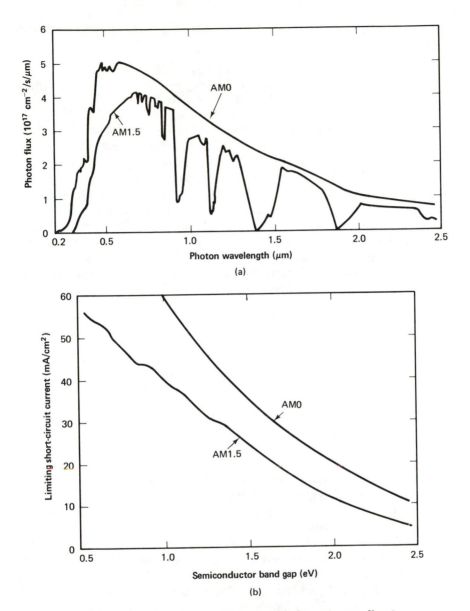

Figure 5.1. (a) Photon flux in sunlight corresponding to the AM0 and AM1.5 energy distributions given in Fig. 1.3. (b) Corresponding upper limits on the short-circuit current density as a function of the energy band gap of the solar cell material.

87

where I_L is the light-generated current and I_0 is the diode saturation current calculated as

$$I_0 = A \left(\frac{q D_e n_i^2}{L_e N_A} + \frac{q D_h n_i^2}{L_h N_D} \right) \tag{5.2}$$

I_0 needs to be as small as possible for maximum V_{oc}. One approach to calculating upper limits on V_{oc} (and hence efficiency) is to assign favorable values to the semiconductor parameters in Eq. (5.2) while still keeping them within the range expected to be required to produce good solar cells (Ref. 5.1). For silicon, this gives a maximum V_{oc} of about 700 mV. The corresponding maximum fill factor is 0.84. Combining with the results of the preceding section for I_{sc} allows the maximum energy-conversion efficiency to be found.

The parameter in Eq. (5.2) which depends most strongly on the choice of semiconductor material is the square of the intrinsic concentration, n_i^2. From Chapter 2,

$$n_i^2 = N_C N_V \exp \left(-\frac{E_g}{kT} \right) \tag{5.3}$$

A reasonable estimate of the minimum value of the saturation current density as a function of band gap from Eq. (5.2) is

$$I_0 = 1.5 \times 10^5 \exp \left(-\frac{E_g}{kT} \right) \qquad \text{A/cm}^2 \tag{5.4}$$

This relationship ensures that the maximum value of V_{oc} decreases with decreasing band gap. This trend is opposite from that observed for I_{sc}. It follows that there will be an optimum-band-gap semiconductor for highest efficiency.

This is demonstrated in Fig. 5.2, which shows maximum energy-conversion efficiency calculated as outlined above as a function of band gap. The peak efficiency occurs for a band gap in the range 1.4 to 1.6 eV and increases from 26 to 29% as the air mass increases from 0 to 1.5. Silicon's band gap is lower than optimum, but maximum efficiencies are still relatively high. GaAs has a near-optimal band gap (1.4 eV).

A major contributor to these relatively low maximum efficiencies is the fact that each photon absorbed creates one electron-hole pair regardless of its energy. The electron and hole quickly relax back to the band edges, emitting phonons (Fig. 5.3). Even though

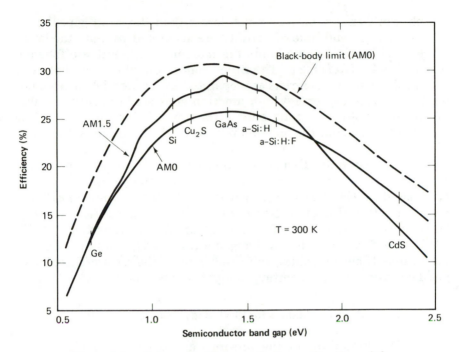

Figure 5.2. Solar cell efficiency limits as a function of the band gap of the cell material. The solid lines are semiempirical limits for AM0 and AM1.5 illumination; the dashed line is based on thermodynamic considerations for blackbody solar cells under AM0 radiation.

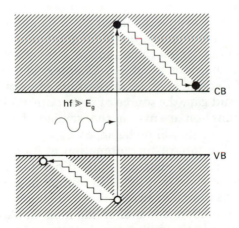

Figure 5.3. One of the major loss mechanisms in solar cells. An electron–hole pair created by a high-energy photon quickly "thermalizes" or relaxes back to the edges of the respective carrier bands. The energy wasted is dissipated as heat.

89

the photon's energy may be much larger than the band gap, the resulting electron and hole effectively are separated by only the latter energy. This effect alone limits the maximum achievable efficiency to about 44% (Ref. 5.2). The other major contributor is that even though the generated carriers are separated by a potential corresponding to the band-gap energy, *p-n* junction cells are inherently capable of giving a voltage output only a fraction of this potential. For example, for silicon, the maximum value of this fraction is 0.7/1.1 ≈ 60%.

The foregoing discussion is limited to the case of an individual cell exposed directly to sunlight. Experimentally, devices of this type based on GaAs have exceeded 20% efficiency. There exist techniques mentioned later whereby photovoltaic system efficiencies can be enhanced further. An efficiency of 28.5% was reported in 1978 (Ref. 5.3) for a system involving multiple cells. Despite the low maximum-efficiency values, solar cells remain the most efficient way yet demonstrated of converting sunlight to electricity.

5.2.4   Efficiency Limits for Black-Body Cells

The limitation of the previous approach to calculating the maximum V_{oc} is its empirical nature. There is a small probability that an unanticipated change in materials technology and/or solar cell design could produce higher efficiency cells than the limits of Fig. 5.2. A more fundamental approach is developed in Ref. 5.2 for the case of a black-body solar cell. Such a body absorbs all radiation incident on it. It is plausible that such a cell would have an efficiency limit at least as high as any non-black-body types.

Black bodies emit radiation with a spectral distribution characteristic of their temperature (Chapter 1). Hence, a black-body solar cell in equilibrium emits photons. For photons of energy larger than the band gap, the source of these photons is predominantly radiative recombination events in the semiconductor. In thermal equilibrium, these events will be balanced by an equal generation rate. A *lower limit* to the rate of recombination at thermal equilibrium is then the total number of photons emitted per unit time with energy larger than the band gap. In a cell with minimum recombination, the recombination rate can be shown to increase exponentially with bias. This leads to an equation identical to the ideal diode law already discussed for solar cells in the dark, with the value of I_0 equal to the electronic charge multiplied by the equilibrium rate of recombination throughout the cell.

Performing the appropriate calculations for silicon gives the minimum value of I_0, which corresponds to a maximum possible V_{oc}

of 850 mV for a black-body silicon solar cell. Calculations for different band-gap semiconductors are shown as the dashed curve in Fig. 5.2. This places the upper efficiency limit for a directly exposed individual cell above 30%.

5.3 EFFECT OF TEMPERATURE

Since the operating temperature of solar cells in the field can vary over wide extremes, it is necessary to understand the effect of temperature on performance.

The short-circuit current of solar cells is not strongly temperature-dependent. It tends to increase slightly with increasing temperature. This can be attributed to increased light absorption, since semiconductor band gaps generally decrease with temperature. The other cell parameters, the open-circuit voltage and the fill factor, both decrease.

The relation between short-circuit current and open-circuit voltage is

$$I_{sc} = I_0 (e^{qV_{oc}/kT} - 1) \tag{5.5}$$

Neglecting the small negative term, this can be written as

$$I_{sc} = A T^\gamma e^{-E_{g0}/kT} e^{qV_{oc}/kT} \tag{5.6}$$

where A is independent of temperature, E_{g0} is the linearly extrapolated zero temperature band gap of the semiconductor making up the cell, and γ includes the temperature dependencies of the remaining parameters determining I_0. Its value generally lies in the range 1 to 4. Differentiating gives, with $V_{g0} = E_{g0}/q$,

$$\frac{dI_{sc}}{dT} = A\gamma T^{\gamma-1} e^{q(V_{oc}-V_{g0})/kT} + A T^\gamma \left(\frac{q}{kT}\right)\left[\frac{dV_{oc}}{dT}\right.$$
$$\left. - \left(\frac{V_{oc} - V_{g0}}{T}\right)\right] e^{q(V_{oc}-V_{g0})/kT} \tag{5.7}$$

Neglecting dI_{sc}/dT in comparison with more significant terms results in the expression

$$\boxed{\frac{dV_{oc}}{dT} = -\frac{V_{g0} - V_{oc} + \gamma(kT/q)}{T}} \tag{5.8}$$

This predicts an approximately linear decrease in V_{oc} with increasing temperature. Substituting values for silicon ($V_{g0} \sim 1.2$ V, $V_{oc} \sim 0.6$ V, $\gamma \sim 3$, $T = 300$ K) gives

$$\frac{dV_{oc}}{dT} = - \frac{1.2 - 0.6 + 0.078}{300} \quad \text{V}/^{\circ}\text{C} \qquad (5.9)$$

$$= - 2.3 \quad \text{mV}/^{\circ}\text{C} \qquad (5.10)$$

This agrees well with the experimental results.[2]   Hence, for silicon, V_{oc} decreases by about 0.4% per $^{\circ}$C. The ideal fill factor depends on the value of V_{oc} normalized to kT/q. Hence, the fill factor also decreases with increasing temperature.

The dominant variation is that of V_{oc}. This causes the power output and efficiency to decrease with increasing temperature. For silicon, the power output decreases by 0.4 to 0.5% per $^{\circ}$C. This dependency is reduced for larger band-gap material. For example, GaAs cells are only about half as sensitive to increasing temperature as are silicon cells.

5.4 EFFICIENCY LOSSES

5.4.1 General

A schematic cross section of an actual *p-n* junction solar cell is shown in Fig. 5.4. Actual devices are significantly less efficient than the ideal limits discussed, owing to various additional loss mechanisms. In later chapters, it will be shown how solar cells are designed to obtain the best trade-off between the loss mechanisms discussed in the following sections.

5.4.2 Short-Circuit Current Losses

There are three types of losses in solar cells which could be described as being of an "optical" nature:

1. It was mentioned in Section 3.2 that bare silicon is quite reflective. The antireflection (AR) coating of Fig. 5.4 reduces such reflection losses to about 10%.

---

[2]This can be attributed largely to the fact that the form of Eq. (5.8) is valid for conditions more general than those for which it was derived here.

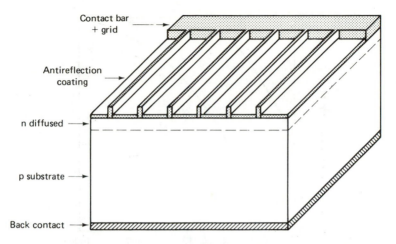

Figure 5.4. Major features of a solar cell. Dimensions in the vertical direction are exaggerated compared to lateral dimensions for the purposes of illustration.

2. The necessity of making electrical contact to both p- and n-type regions of solar cells generally results in a metal grid contact on the side of the cell exposed to sunlight. This blocks 5 to 15% of the incoming light.

3. Finally, if the cell is not thick enough, some of the light of appropriate energy that does get coupled into the cell will pass straight out the back. This determines the minimum thickness of semiconductor material required. Indirect-band-gap semiconductors require more material than direct-gap materials, as indicated by the calculated results for Si and GaAs in Fig. 5.5.

Another source of I_{sc} loss is recombination in the bulk semiconductor and at surfaces. It was indicated in Chapter 4 that only electron–hole pairs generated near the p-n junction itself contribute to I_{sc}. Carriers generated well away from the junction have a high probability of recombining before they complete the trip from the point of generation to the device terminals.

5.4.3 Open-Circuit Voltage Losses

The fundamental process determining V_{oc} is recombination in the semiconductor. This is brought out by the approach to calculating limits on V_{oc} in Section 5.2.4 The lower the recombination

93

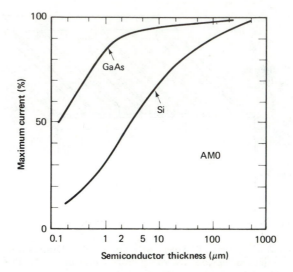

Figure 5.5. Effect of cell thickness on the percentage of full short-circuit current that may ideally be generated by a solar cell. Note the difference between the direct-band-gap semiconductor (GaAs) and indirect-band-gap silicon.

rate in the semiconductor, the higher is V_{oc}. Both bulk and surface recombination are important.

One important effect that can limit V_{oc} is recombination through trapping levels in depletion regions. This recombination mechanism is particularly effective in such regions. Referring back to the expression for this mechanism (Chapter 3), we obtain

$$U = \frac{np - n_i^2}{\tau_{h0}(n + n_1) + \tau_{e0}(p + p_1)} \qquad (5.11)$$

This rate will have its peak value when both n_1 and p_1 are small and both n and p are small. Both these conditions occur for traps near midgap located within the depletion region. In the analysis of the dark characteristics of *p-n* diodes in Chapter 4, recombination in depletion regions was neglected on the basis of the depletion-layer width, W, being very small (*approximation 5*). However, the enhanced value of recombination in this region can make it quite important in some situations.

Including such depletion-region recombination adds an additional term to the dark current–voltage characteristics, which then become

$$I = I_0(e^{qV/kT} - 1) + I_W(e^{qV/2kT} - 1) \qquad (5.12)$$

94

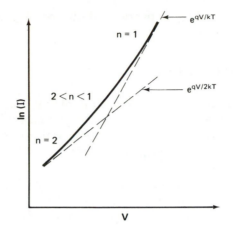

Figure 5.6. Semilogarithmic plots of the dark current–voltage characteristics of a *p-n* junction diode in the dark, including the effect of recombination in the depletion region.

where I_0 has the same value as before and I_W has the value (Ref. 5.4)

$$I_W = \frac{qAn_i\pi}{2\sqrt{\tau_{e0}\tau_{h0}}} \frac{kT}{q\,\xi_{max}} \tag{5.13}$$

where ξ_{max} is the maximum electric field strength in the junction. Its value is given by Eq. (4.4) for junctions uniformly doped on each side.

These characteristics are plotted on a semilogarithmic plot in Fig. 5.6. The second term of Eq. (5.12) dominates at low currents, the first at high currents.

It is possible to write Eq. (5.12) in the form

$$I = I_0'(e^{qV/nkT} - 1) \tag{5.14}$$

where n is known as the *ideality factor*. It varies with current level as does I_0'. From Eq. (5.12), n decreases from 2 at low currents to 1 at higher currents.[3] Since the illuminated solar cell characteristics are those of Fig. 5.6 shifted down into the fourth quadrant, it can be seen that the presence of this additional depletion-region recombination current can act to decrease V_{oc}.

[3] An additional region where n again approaches 2 can be obtained at high currents when minority carrier concentrations approach those of the majority carriers in some regions of the device (Ref. 5.5).

Other technology-dependent limits on V_{oc} are discussed in later chapters.

5.4.4 Fill Factor Losses

Recombination in the depletion region can also reduce the fill factor. If the ideality factor, n, of the preceding section is greater than unity, the fill factor is that calculated for the ideal case (Fig. 4.12) at a voltage V_{oc}/n. This will give a lower value than when n equals unity.

Defining the normalized voltage, v_{oc}, as $V_{oc}/(nkT/q)$ in this more general case, the empirical expression for the fill factor given in Chapter 4 remains valid with an accuracy of about four significant digits for $v_{oc} > 10$:

$$FF_0 = \frac{v_{oc} - \ln(v_{oc} + 0.72)}{v_{oc} + 1} \tag{5.15}$$

Solar cells generally have a parasitic series and shunt resistance associated with them, as indicated in the solar cell equivalent circuit of Fig. 5.7. There are several physical mechanisms responsible for these resistances. The major contributors to the series resistance, R_S, are the bulk resistance of the semiconductor material making up the cell, the bulk resistance of the metallic contacts and interconnections, and the *contact resistance* between the metallic contacts and the semiconductor. The shunt resistance, R_{SH}, is caused by leakage across the p-n junction around the edge of the cell and in nonperipheral regions in the presence of crystal defects and precipitates of foreign impurities in the junction region. Both types of parasitic resistances act to reduce the fill factor, as indicated in Fig. 5.8. Very high values of R_S and very low values of R_{SH} can also reduce I_{sc} and V_{oc}, respectively, as indicated.

The magnitude of the effect of R_S and R_{SH} on the fill factor can be found by comparing their values to the *characteristic resistance*

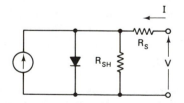

Figure 5.7.   Equivalent circuit of a solar cell.

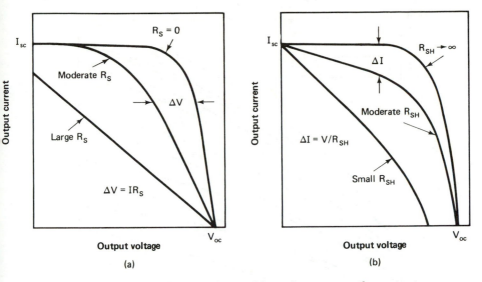

Figure 5.8. Effect of parasitic resistances on the output characteristics of solar cells:

(a) Effect of series resistance, R_S.
(b) Effect of a shunt resistance, R_{SH}.

of a solar cell defined as (Ref. 5.6)

$$R_{CH} = \frac{V_{oc}}{I_{sc}} \tag{5.16}$$

If R_S is a lot less than this quantity or R_{SH} a lot larger, there will be little effect upon the fill factor. Defining a normalized resistance, r_s, as R_S/R_{CH}, an approximate expression for the fill factor in the presence of series resistance is (precise values are given in Fig. 5.9)

$$FF = FF_0(1 - r_s) \tag{5.17}$$

where FF_0 represents the ideal fill factors in the absence of parasitic resistance as approximated quite closely by Eq. (5.15). This expression is accurate to close to two significant digits for $v_{oc} > 10$ and $r_s < 0.4$. Defining a normalized shunt resistance, r_{sh}, as R_{SH}/R_{CH}, a corresponding expression for the effect of shunt resistance, also involving the normalized voltage $v_{oc} = V_{oc}/(nkT/q)$, is (exact values are again given in Fig. 5.9)

$$FF = FF_0 \left\{ 1 - \frac{(v_{oc} + 0.7)}{v_{oc}} \frac{FF_0}{r_{sh}} \right\} \tag{5.18}$$

97

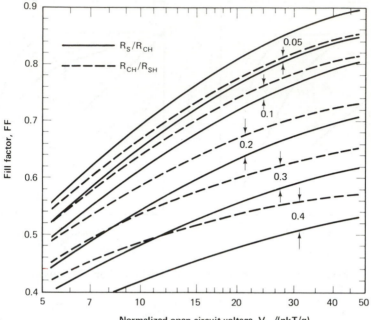

Figure 5.9. General curves for solar cell fill factors as a function of normalized open-circuit voltage. The solid curves show the fill factor as a function of normalized series resistance, R_S/R_{CH}, where $R_{CH} = V_{oc}/I_{sc}$. The dashed curves show the effect of shunt resistance. The normalized parameter in this case is R_{CH}/R_{SH}. These curves allow the fill factor to be found for any combination of open-circuit voltage, temperature, and ideality factor, together with any value of *either* series *or* shunt resistance.

This expression is accurate to close to three significant digits for $v_{oc} > 10$ and $r_{sh} > 2.5$. When both series and shunt parasitic resistance is important, an approximate expression for the fill factor over a more limited range of parameters is Eq. (5.18) with FF_0 replaced by FF as calculated from Eq. (5.17).

5.5 EFFICIENCY MEASUREMENT

It may seem a relatively simple matter to measure solar cell efficiency by measuring the power in the incident sunlight using a pyranometer and the electrical power the cell is generating at the maximum power point. The difficulty with this approach is that the performance of the cell measured in this way will depend greatly on the precise spectral content of the sunlight, which varies with air mass, water-vapor content, turbidity, and so on. Combined with typical uncertainties in pyranometer calibration ($\sim \pm 5\%$), this approach would make comparison difficult between the performance of devices measured at other than the same time and place.

An alternative approach is to use a method based on calibrated reference cells. A central test authority calibrates reference cells under standard illumination conditions. The performance of the cell under test is then measured relative to the reference cell. For this technique to be accurate, two conditions must be satisfied:

1. The response of the reference and test cells to different wavelengths of light (*spectral response*) must be similar within specified limits.

2. The spectral content of the illumination source used for the comparative testing must approximate that of the standard illumination, again within specified limits.

The first condition usually dictates that reference and test cells be made from the same semiconductor material using a similar processing technique. With both conditions satisfied, all measurements are referenced to the standard illumination conditions at the calibration center.

A technique similar to that outlined above has been used in the photovoltaic program of the U.S. Department of Energy (Ref. 5.7). In this case the standard sunlight distribution to which measurements are referenced is the AM1.5 distribution of Fig. 1.3. Recommended illumination sources for testing are natural sunlight (with certain constraints on the presence of clouds, air mass, and the rate of variation of intensity), an appropriately filtered xenon lamp, or an ELH lamp. The latter is an inexpensive tungsten-filament projector lamp incorporating a wavelength-sensitive reflector which allows infrared wavelengths to pass out the rear of the lamp. This enhances the visible content of the output beam, whose spectral content is then reasonably close to sunlight. The illumination source must give a collimated beam of uniform intensity at the test plane and be stable during the time of measurement, all within specified limits.

A typical experimental arrangement for measuring solar cell output characteristics is shown in Fig. 5.10(a). A *four-point contacting* scheme is desirable in which voltage and current leads to the cell under test are kept separate. This eliminates effects due to the series resistance of the test leads and associated contact resistances. The cells are mounted on a temperature-controlled block. Both 25°C and

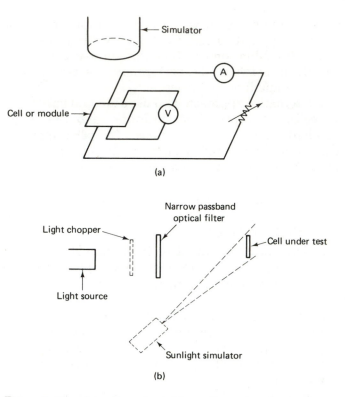

Figure 5.10. (a) Experimental configuration for testing solar cells and modules. (b) Possible setup for measuring the spectral response. The bias light source together with chopped monochromatic light is necessary for cells with a nonlinear spectral response.

28°C are standard temperatures for solar cell measurements. The lamp intensity is adjusted to give the desired intensity as measured by a reference cell. By varying the load resistance, the characteristics of the test cell can then be measured.

The *spectral response* of a test cell can also be measured by direct comparison with the output of a cell with calibrated spectral response. The simplest technique is to use a steady-state source of monochromatic light from a monochromater or that obtained by passing white light through narrow-band optical filters as in Fig. 5.10(b). Since the response of cells to light of increasing intensity may not always be linear, a preferred method is to use a white light source approximating sunlight to bias the cell and measure the incremental response to a small superimposed alternating component of monochromatic light.

5.6 SUMMARY

The upper limit to the efficiency of solar cells lies in the range 26 to 29% and occurs for band gaps in the range 1.4 to 1.6 eV. Several factors combine to ensure that actual solar cells have efficiencies somewhat below ideal. Some relate to coupling the light into the cells, others to excessive recombination in the semiconductor bulk and at surfaces, and still others are due to parasitic resistive effects.

Solar cell efficiencies decrease with increasing temperature, primarily due to the temperature sensitivity of the open-circuit voltage.

Preferred experimental methods for measuring solar cell performance are based on the use of calibrated reference cells to eliminate variables associated with more direct techniques.

EXERCISES

5.1. (a) A solar cell is uniformly illuminated by monochromatic light of wavelength 700 nm and intensity 20 mW/cm^2. What is the corresponding incident photon flux and the upper limit to the short-circuit current output of the cell if its band gap is 1.4 eV?

(b) What would be the corresponding current if the band gap were 2.0 eV?

5.2. When illuminated by the AM1.5 radiation of Table 1.1, the maximum current density obtainable in a given solar cell design is 40 mA/cm^2. If the maximum open-circuit voltage at 300 K is 0.5 V, what is the upper limit to cell efficiency at this temperature?

5.3. A solar cell is uniformly illuminated by monochromatic light of intensity 100 mW/cm^2. The minimum value of cell saturation current density at 300 K is 10^{-11} A/cm^2. Calculate upper limits to the efficiency of conversion of this light to electrical power at this temperature if the wavelength of light is: (a) 450 nm; (b) 900 nm. Assume that the photon energy is larger than the band gap in each case. (c) Explain the difference in the efficiency values obtained.

5.4. Calculate and sketch the upper limit to the *spectral sensitivity* (short-circuit current/power in incident monochromatic light) as a function of wavelength for a silicon solar cell.

5.5. Silicon cells typically have an open-circuit voltage of 0.6 V and GaAs cells, about 1.0 V. Compare the theoretical temperature dependencies of open-circuit voltages for these cells at 300 K on both an absolute and a relative basis. (The effective zero temperature band gaps are 1.2 and 1.57 eV, respectively.)

5.6. From Fig. 5.5, compare the thickness of Si and GaAs required to enable 75% of maximum current output to be obtained under AM0 illumination.

5.7. A solar cell has nearly ideal characteristics, with the ideality factor equal to 1. A second cell has characteristics dominated by recombination in the depletion region and an ideality factor of 2. If both give an open-circuit voltage of 0.6 V at 300 K, compare their ideal fill factors.

5.8. A solar cell at 300 K has an open-circuit voltage of 550 mV, a short-circuit current of 2 A, and an ideality factor of 1.3. Calculate the fill factor under the following conditions: (a) series resistance is 0.08 Ω, shunt resistance large. (b) series resistance negligible, shunt resistance is 1 Ω. (c) series resistance is 0.08 Ω, shunt resistance is 2 Ω. (d) series resistance is 0.02 Ω, shunt resistance is 1 Ω.

REFERENCES

[5.1]   H. J. HOVEL, *Solar Cells*, Vol. 11, Semiconductors and Semimetals Series (New York: Academic Press, 1975).

[5.2]   W. SHOCKLEY AND H. J. QUEISSER, "Detailed Balance Limit of Efficiency of *p-n* Junction Solar Cells," *Journal of Applied Physics 32* (1961), 510–519.

[5.3]   R. L. MOON et al., "Multigap Solar Cell Requirements and the Performance of AlGaAs and Si Cells in Concentrated Sunlight," *Conference Record, 13th IEEE Photovoltaic Specialists Conference*, Washington, D.C. (1978), pp. 859–867.

[5.4]   C. T. SAH et al., "Carrier Generation and Recombination in *p-n* Junctions. . . ," *Proceedings of the IRE 45* (1957), 1228–1243.

[5.5]   J. G. FOSSUM et al., "Physics Underlying the Performance of Back-Surface-Field Solar Cells," *IEEE Transactions on Electron Devices ED-27* (1980), 785–791.

[5.6]   M. A. GREEN, "General Solar Cell Factors. . . ," *Solid-State Electronics 20* (1977), 265–266.

[5.7]   *Terrestrial Photovoltaic Measurement Procedures*, Report ERDA/NASA/1022-77/16, June 1977.

STANDARD SILICON
SOLAR CELL TECHNOLOGY

6.1 INTRODUCTION

After the development of the first reasonably efficient silicon cells in 1953, these cells found a major application in generating electricity for spacecraft. The first such use was in Vanguard I in 1958. Since then, a small but growing industry has developed to supply cells for the ever-increasing number of communication satellites and other space craft. Stringent requirements on performance and reliability led to the development of a standard cell-processing sequence which remained virtually unchanged throughout the 1960s and early 1970s.

Since 1973, increased interest in renewable energy resources has led to several companies currently producing cells specifically for terrestrial use. A list of several solar cell manufacturers is given in Table 6.1. Initially these tended to carry over the standard technology developed for space cells. Although different requirements for terrestrial use have since produced some significant changes in the techniques used to fabricate cells, the standard technology will be described in this chapter to provide a base for discussing these changes and probable future developments.

Table 6.1. MANUFACTURERS OF SOLAR CELLS (circa 1980)

United States

Applied Solar Energy Corporation 15251 East Don Julian Road City of Industry, CA 91746	Solar Power Corporation 20 Cabot Road Woburn, MA 01801
Arco Solar, Inc. 20554 Plummer Street Chatsworth, CA 91311	Solarex Corporation 1335 Piccard Drive Rockville, MD 20850
Motorola, Inc. Solar Energy Department Phoenix, AZ 95008	Solec International, Inc. 12533 Chadron Avenue Hawthorne, CA 90250
Photon Power, Inc. 10767 Gateway West El Paso, TX 79935	Solenergy Corporation 23 North Avenue Wakefield, MA 01880
Photowatt International, Inc. 21012 Lassen Street Chatsworth, CA 91311	Spectrolab, Inc. 12484 Gladstone Avenue Sylmar, CA 91342
SES Incorporated Tralee Industrial Park Newark, DE 19711	Spire Corporation Patriots Park Bedford, MA 01730

Europe

AEG Telefunken Discrete Component Dept. P.O. Box 1109 7100 Helibronn, W. Germany	RTC (Philips Group) Route de la Delivrande 14000 Caen-Cedex, France

Japan

Japan Solar Energy Co. 11-17 Kogahonmachi Fushimiku, Kyoto Matsushita Electric Kadoma, Osaka	Sharp Corp. Engineering Division 2613.1 Ichinomoto Tenri-Shi, Nara

Australia	*India*
Tideland Energy Pty. Ltd. P.O. Box 519 Brookvale, N.S.W. 2100	Central Electronics Ltd. Site 4, Industrial Area Sahibabad, U.P. 201005

The standard technology for making cells can be broken down into the following stages:

1. Reduction of sand to metallurgical-grade silicon.
2. Purification of metallurgical-grade silicon to semiconductor-grade silicon.
3. Conversion of semiconductor-grade silicon to single-crystal silicon wafers.
4. Processing of single-crystal silicon wafers into solar cells.
5. Solar cell encapsulation into weatherproof solar cell modules.

6.2 SAND TO METALLURGICAL-GRADE SILICON

Silicon is the second most abundant element in the earth's crust. The source material for the extraction of silicon is silicon dioxide, the major constituent of sand. However, in the present commercial extraction process, the crystalline form of silicon dioxide, quartzite, is used. This material is reduced in large arc furnaces, as illustrated in Fig. 6.1, by carbon (in the form of a mixture of wood chips, coke,

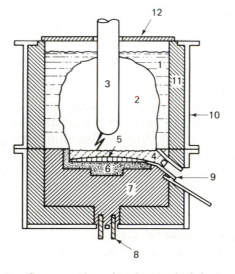

Figure 6.1. Cross section showing typical features of an arc furnace as used to produce metallurgical-grade silicon: 1, carbon and quartz; 2, cavity; 3, electrode; 4, silicon; 5, silicon carbide; 6, hearth; 7, electrode paste, 8, copper electrodes; 9, tapping spout; 10, cast-iron shell; 11, ceramic; 12, graphite lid. (After Ref. 6.1, © 1976 IEEE.)

Table 6.2. TYPICAL CONCENTRATIONS OF IMPURITIES IN
METALLURGICAL-GRADE SILICON

Impurity	Concentration range (parts per million, atomic)
Al	1500–4000
B	40–80
Cr	50–200
Fe	2000–3000
Mn	70–100
Ni	30–90
P	20–50
Ti	160–250
V	80–200

and coal) to produce silicon according to the reaction (Ref. 6.1)

$$SiO_2 + 2C \longrightarrow Si + 2CO \qquad (6.1)$$

Silicon is periodically poured from the furnace and blown with oxygen or oxygen/chlorine mixtures to further purify it. It is then poured into shallow troughs, where it solidifies and is subsequently broken into chunks.

Of the order of 1 million metric tons of this metallurgical-grade silicon (MG-Si) are produced globally each year, primarily for use in the steel and aluminum industries. It is generally 98 to 99% pure, with the major impurities being iron and aluminum, as indicated by the typical breakdown of Table 6.2. The reduction process is reasonably energy-efficient. The total processing energy requirements are similar to those required for related metals such as aluminum or titanium. The material is also quite inexpensive. A very small portion of the total MG-Si production is further purified to *semiconductor grade* (SeG) for the electronics industry, which uses few thousand metric tons of this material annually.

6.3 METALLURGICAL-GRADE SILICON TO SEMICONDUCTOR-GRADE SILICON

For use in solar cells as well as other semiconductor devices, silicon must be much purer than metallurgical grade. The standard approach to purifying it is known as the Siemens process (Ref. 6.2). The MG-Si is converted to a volatile compound that is condensed and refined by fractional distillation. Ultrapure silicon is then extracted from this refined product.

The detailed processing sequence is that a bed of fine MG-Si particles is fluidized with HCl in the presence of a Cu catalyst to promote the reaction

$$Si + 3HCl \longrightarrow SiHCl_3 + H_2 \qquad (6.2)$$

The gases emitted are passed through a condenser and the resulting liquid subjected to multiple fractional distillation to produce SeG-SiHCl$_3$ (trichlorosilane). This is a source material for the silicone industry.

To extract SeG-Si, the SeG-SiHCl$_3$ is reduced by hydrogen when mixtures of the gases are heated. Silicon is deposited in a fine-grained polycrystalline form onto an electrically heated silicon rod according to the reaction

$$SiHCl_3 + H_2 \longrightarrow Si + 3HCl \qquad (6.3)$$

The latter step not only requires a lot of energy but also has a low yield ($\sim 37\%$). This is the main reason for the massive increase in the energy required to produce SeG-Si compared to MG-Si discussed in Section 6.7. The cost increase during this transition is even larger. Consequently, the purification of MG-Si has been a prime target for improved technology.

6.4   SEMICONDUCTOR-GRADE POLYSILICON TO SINGLE-CRYSTAL WAFERS

For the semiconductor electronics industry, silicon must not only be very pure, but it must also be in single-crystal form with essentially zero defects in the crystal structure. The major method used to produce such material commercially is the *Czochralski process* illustrated in Fig. 6.2. The SeG polycrystalline silicon is melted in a crucible with trace levels of one of the dopants required in the completed device added. For solar cells, boron, a *p*-type dopant, is normally used. Using a seed crystal and with very close temperature control, it is possible to pull from the melt a large cylindrical single crystal of silicon. Crystals of diameter in excess of 12.5 cm and 1 to 2 m in length are routinely grown in this manner.

As seen in Section 5.4.2, silicon solar cells need only be 100 μm or so thick to absorb most of the appropriate wavelengths in

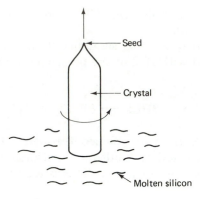

Figure 6.2. Essential features of the Czochralski process for the growth of large cylindrical ingots of single-crystal silicon.

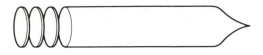

Figure 6.3. Slicing of thin wafers from a cylindrical ingot. The techniques used for this slicing process are described and compared in Ref. 6.3. About half the ingot is wasted as kerf or cutting loss in this process.

sunlight. Therefore, the large single crystal is sliced up into wafers which are as thin as possible, as indicated in Fig. 6.3. With present wafering technology (Ref. 6.3), it is difficult to cut wafers from the large crystals previously described which are any thinner than 300 μm and still retain reasonable yields. More than half the silicon is wasted as kerf or cutting loss in the process. The low overall yield in going from SeG-Si to single-crystal wafers is another weak link in the standard silicon technology.

6.5 SINGLE-CRYSTAL WAFERS TO SOLAR CELLS

After etching the silicon wafers (to remove damage from the wafering process) and cleaning them, additional impurities are introduced into the cell in a controlled manner by a high-temperature impurity diffusion process.

It was mentioned in the preceding section that, in standard solar cell technology, boron was normally added to the melt in the Czochralski process which then results in p-type wafers. To form

108

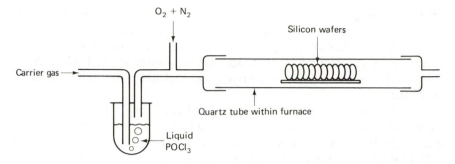

Figure 6.4. Phosphorus diffusion process.

a solar cell, n-type impurities must be introduced to give a p-n junction. Phosphorus is the impurity generally used. In the most common process, a carrier gas is bubbled through phosphorus oxychloride ($POCl_3$), mixed with a small amount of oxygen, and passed down a heated furnace tube in which the wafers are stacked as illustrated in Fig. 6.4. This grows an oxide layer on the surface of the wafers containing phosphorus. At the temperatures involved (800 to 900°C), the phosphorus diffuses from the oxide into the silicon. After about 20 min, the phosphorus impurities override the boron impurities in the regions near the surface of the wafer to give a thin, heavily doped n-type region shown in Fig. 6.5(a). In subsequent processing, the oxide layer is removed, as are the junctions at the side and back of the cell to give the structure of Fig. 6.5(b).

Metal contacts are then attached to both the n-type and the p-type regions. In the standard technology, the process used is known as *vacuum evaporation*. The metal to be deposited is heated in a vacuum to a high-enough temperature to cause it to melt and vaporize. It will then condense on any cooler parts of the vacuum system in direct line of sight, including the solar cells. The back contact is normally deposited over the entire back surface, while the top

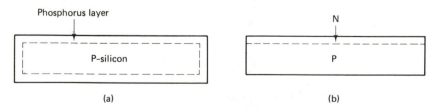

Figure 6.5. Distribution of phosphorus impurities:
(a) Immediately after diffusion process.
(b) After etching of back and side of wafer.

109

contact is required in the form of a grid. Two techniques are available for defining such a pattern. One is to use a metal *shadow mask* (Fig. 6.6). Alternatively, the metal can be deposited over the entire front surface of the cell and subsequently etched away from unwanted regions using a photographic technique known as *photolithography*.

The contact is usually made up of the three separate layers of metal. A thin layer of titanium is used as the bottom layer for good adherence to silicon. The top layer is silver for low resistance and solderability. Sandwiched between these two is a layer of palladium which prevents an undesirable reaction between the Ti and Ag layers in the presence of moisture. After deposition, the contacts are sintered at 500 to 600°C to give good adherence and low contact resistance. Finally, a thin antireflection (AR) coating is deposited on the top of the cell by the same vacuum evaporation process.

Yields of about 90% from starting wafers to completed terrestrial cells can be obtained. All the processes above are done on a *batch* basis, where a batch of perhaps 40 to 100 wafers moves through the processing sequence at the same time. This makes the processing very labor-intensive. In addition, the vacuum evaporation equipment is expensive compared to its throughput. Moreover,

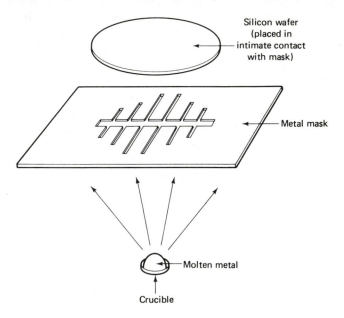

Figure 6.6. Essential features of vacuum evaporation of metallic layers as well as the use of a metal mask to define the top metal grid pattern.

because of the nature of the evaporation process, only a small fraction of the material being evaporated ends up where desired. This is quite extravagant when dealing with material as expensive as silver.

6.6 SOLAR CELLS TO SOLAR CELL MODULES

6.6.1 Module Construction

Solar cells require encapsulation not only for mechanical protection but also to provide electrical isolation and a degree of chemical protection. Encapsulation provides mechanical rigidity to support the brittle cells and their flexible interconnections. It also provides protection from mechanical damage as may be caused by hail, birds, and objects dropped or thrown onto the module. Encapsulation protects metallic contacts and interconnections from corrosive elements in the atmosphere. Finally, it provides electrical isolation of the voltages generated by the panel. These may reach voltages as high as 1500 V above ground in some systems. The durability of the encapsulation will determine the ultimate operating life of the module, which ideally should be 20 years or more.

Additional features an encapsulation scheme must possess are ultraviolet (UV) stability, tolerance to temperature extremes and thermal shocks without stressing cells to fracture, resistance to abrasion as might occur during dust storms, self-cleaning ability, ability to keep cell temperature low to minimize power loss, and low cost.

Several different approaches to module design are possible. One essential component is a structural layer to provide rigidity. This layer can be either at the back or front of the module as indicated in Fig. 6.7. Cells can be either bonded directly to this layer and encapsulated in a flexible pottant or enclosed in a laminate supported by it. The final layer, if at the rear of the module, acts as a moisture barrier. If at the top, it has the additional duties of imparting self-cleaning properties and improving impact resistance. Some form of moisture seal is incorporated around the edges of the module.

For the *structural back* configuration of Fig. 6.7(a) and (b), the most common materials for the structural back have been anodized aluminum, porcelainized steel, epoxy board, or window glass. Wood composites such as chipboard may prove to be the cheapest overall possibilities for this layer. For the *structural front configuration* of Fig. 6.7(c) and (d), glass is the obvious choice for the structural layer. It combines excellent weatherability with low cost and good self-cleaning properties. Most designs use a toughened or tem-

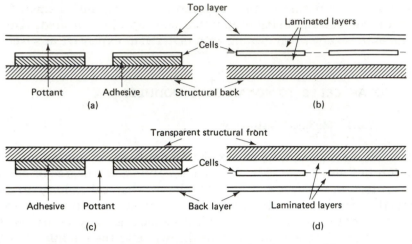

Figure 6.7. Schematic diagrams showing possible approaches to the encapsulation of solar cells:
(a)   Structural back configuration.
(b)   Structural back configuration with laminated layers.
(c)   Structural front configuration.
(d)   Structural front configuration with laminated layers.

pered glass with a low iron content for good light transmission. For the adhesive and pottant layers, silicones have been widely used. They have good UV stability, low light absorption, and are appropriately elastic to alleviate thermal stresses in the module, but they are expensive. Polyvinyl butyral (PVB) and ethylene/vinyl acetate (EVA) have also been used by several manufacturers for the corresponding layers in the laminated approach.

The top layer of the *structural back configuration* serves to impart self-cleaning properties and in some designs acts as a moisture barrier. Low-iron glass again has been a popular choice. Polymers such as acrylics have also been used. Some manufacturers have concentrated on producing a moisture-resistant cell, thereby relaxing encapsulation requirements. These may use a soft silicone encapsulant and a harder silicone overlying layer to improve self-cleaning. The most popular choices for the back layer in the structural front approach have been Mylar or Tedlar to act as a moisture barrier. However, all polymers are permeable to moisture to some extent. A solution to this has been to use a thin foil of aluminum or stainless steel embedded between layers of an appropriate polymer.

If the back layer is white, it is possible to increase the module output marginally using the *zero-depth concentration effect* (Ref. 6.4). Some of the light striking regions of the module between cells

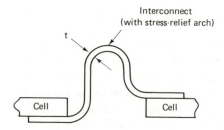

Figure 6.8. Stress-relief loop, as generally required in the metallic interconnections between cells to prevent fatigue due to cyclic thermal and wind loading stresses. For maximum effect, the thickness of the interconnect, t, should be small and the height of the arch high as discussed in Ref. 6.5.

will be scattered by the back layer and channeled by the glass super-strate to active regions of the module. This can give a boost in the module output, particularly when the packing density of cells is low.

Another important area of module design concerns the interconnections between the cells. It is common practice to use multiple interconnects for redundancy. This increases the module tolerance to interconnect failure (by corrosion or fatigue) as well as to cracked cells. Cyclic stresses are set up in the interconnects due to differentials in temperature expansion coefficients and wind loading. A stress-relief loop as indicated in Fig. 6.8 is generally required in the interconnect (Ref. 6.5).

6.6.2 Cell Operating Temperature

Different module designs will cause cells encapsulated within them to reach different temperatures under identical operating conditions. Since the cell performance is adversely affected by increasing temperature (Section 5.3), the modules operating at cooler temperature will have a relative performance boost.

Rather than comparing the performance of different modules at the same temperature, it is more appropriate to compare the performance at different temperatures. In each case, this would be the temperature that the cells would reach under typical operating conditions. If a standard set of operating conditions (insolation level, wind velocity and direction, ambient temperature, electrical loading of the cells) is defined, it is possible to associate a particular temperature, the *nominal operating cell temperature* (NOCT), with each module type. Experimental techniques have been developed (Ref.

113

6.6) for calculating this temperature from field data at nonstandard operating conditions.

Field data indicate that the operating temperature of the solar cell above ambient is roughly proportional to the intensity of the incident sunlight, provided that wind velocity is not excessive. As a rule of thumb, most commercial modules allow the cell temperature to rise about 30°C above ambient under full sun irradiation (100 mW/cm²) when mounted on an open frame. Hence, an approximate expression for the cell temperature is

$$T_{cell}(^{\circ}C) = T_{ambient}(^{\circ}C) + 0.3 \times \text{intensity (in mW/cm}^2) \qquad (6.4)$$

When roof-mounted, cell operating temperatures would tend to be higher.

6.6.3   Module Durability

The durability of the solar cell encapsulation will ultimately determine the operating life of a solar cell system, since there is no "wear-out" mechanism associated with cell operation in the terrestrial environment.

The types of module degradation that have been observed in the field in the past have been:

1. Breakage of cells due to excessive mechanical stress caused by thermal fluctuations or, more directly, by hail damage or vandalism.
2. Corrosion of metallization.
3. Delamination of different layers of the encapsulation.
4. Discoloring of encapsulating material.
5. Accumulation of dirt on modules with "soft" top surfaces.
6. Breakage of interconnects due to inadequate stress relief.

With increasing field experience, module design has steadily improved to the stage where a 20-year operating life is feasible. Accelerated life-testing of new module designs is normally carried out by subjecting them to the types of stress below:

1. Thermal cycling
2. High humidity

3. Prolonged ultraviolet radiation
4. Cyclic pressure loading

Combinations of these stresses will often enhance module degradation. Other types of qualification testing that may be carried out include:

1. Impact testing
2. Abrasion resistance
3. Self-cleaning properties
4. Flexibility (to test for mounting onto warped surfaces)
5. Electrical insulation (particularly after accelerated life-testing)

Although dirt accumulation has caused severe degradation in some locations for modules with soft top surfaces (Ref. 6.7), this is not a major problem with glass-faced modules. Self-cleaning by rain and wind keeps the power loss from this effect to less than 10%. Since cells can operate from scattered light, modules have been found to give a reasonable percentage of their peak output even when deliberately covered with sufficient dust so that individual cells in the module can barely be distinguished (Ref. 6.8).

6.6.4 Module Circuit Design

The electrical circuit aspects of the way solar cells are interconnected within solar modules can have a substantial effect on the field performance and operating life of the modules.

When solar cells are connected together, mismatches in their operating characteristics ensure that the output power of the combination is less than the sum of the maximum output power of the constituents. The difference, the *mismatch loss*, is the most significant for cells connected in series.

Even more important than the loss in power is the potential for overheating in the poorest cell of a series string. Figure 6.9 shows the output characteristic of the lowest current output cell in a series string together with the combined output of the rest of the cells. The voltage across each of these components must be equal and opposite in sign when the module is short-circuited. The module short-circuit current can be found by reflecting one of the curves of Fig. 6.9 in the current axis as shown and finding its intercept with the other. Note that, in this condition, the poorest cell is reversed-biased

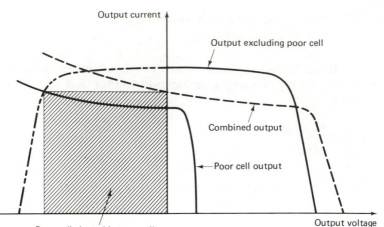

Output current

Output excluding poor cell

Combined output

←—Poor cell output

Output voltage

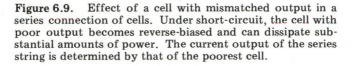

Power dissipated in poor cell
when combination is short-circuited

Figure 6.9. Effect of a cell with mismatched output in a series connection of cells. Under short-circuit, the cell with poor output becomes reverse-biased and can dissipate substantial amounts of power. The current output of the series string is determined by that of the poorest cell.

and power equal to the shaded area shown is being dissipated in it. It is apparent from Fig. 6.9 that power up to the maximum generating capability of the rest of the series string can be dissipated in the poorest cell under some conditions. Such effects can cause an excessive temperature rise in localized regions of the poorest cell, which can damage the cell encapsulation and cause the eventual deterioration of an entire module. The same effect can be caused by partial shadowing of some cells in a module or by cracked cells.

It is not difficult to show that the open-circuit voltage of a series string is the sum of those of the individual cells. Another property apparent from Fig. 6.9 is that the short-circuit current will be determined by that of the lowest output current cell in the string. It follows that severe mismatch in the short-circuit currents can cause the current-generating capabilities of the better cells in the string to be entirely wasted. Although similar *mismatch losses* also occur in parallel connections of cells, they are not nearly as severe.

Two techniques are available for lessening the severity of the foregoing effects (Ref. 6.9). One is a technique known as series-paralleling and the other is by the use of bypass diodes. Figure 6.10 defines the terminology used in describing such module circuit design techniques.

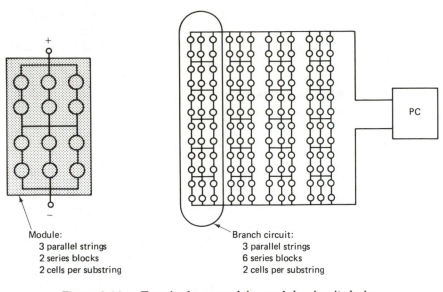

Module:
3 parallel strings
2 series blocks
2 cells per substring

Branch circuit:
3 parallel strings
6 series blocks
2 cells per substring

Figure 6.10. Terminology used in module circuit design. The block labeled PC is power conditioning equipment. (After Ref. 6.9, © 1980 IEEE.)

By increasing the numbers of *series blocks* and *parallel strings* per module or *branch circuit*, tolerance to cell mismatch, cracked cells, and partial shadowing is increased. This is the technique known as *series-paralleling*. An alternative approach is to use a *bypass diode* connected across one or more *series blocks* in the module. The by-pass diode becomes forward-biased when a series block becomes reversed-biased. This limits the power dissipated in this block as well as providing a low-resistance path for the module or *branch string* current.

6.7 ENERGY ACCOUNTING

It is naturally important that any power-generating device used on a large scale delivers more energy over its operating life than that invested in constructing, commissioning, and maintaining it. How do silicon cells fabricated using the standard technology described in this chapter fare in this regard?

The extraction of metallurgical-grade silicon from quartzite is quite an energy efficient process. Taking into account the energy required to mine, deliver, and prepare the raw materials for this

117

process together with the processing energy required, the equivalent of about 24 kWh of electrical energy [kWh(e)] is required to produce 1 kg of this material (Ref. 6.1). This is roughly in the same area as that for aluminum [19 kWh(e)/kg] or titanium [46 kWh(e)/kg] calculated on the same basis.

The Siemens process for purifying this to semiconductor-grade quality is costly, inefficient, and energy-intensive. This has made it an obvious target for replacement in future silicon cell technology. On the same basis as above, the energy content of semiconductor-grade silicon is 621 kWh(e)/kg (Ref. 6.1).

The conversion of this pure silicon to single-crystal sheet form via the Czochralski process and subsequent slicing of the cylindrical ingots produced by this process does not make good use of the semiconductor-grade silicon. Sheet is produced at the rate of about 0.4 m^2/kg of this material. The major reason for this low yield is the nonideal nature of the cutting process, which wastes half the starting material and produces wafers thicker than required from photovoltaic considerations. The energy content of silicon wafers on the same basis as before is 1700 kWh(e)/m^2.

Cell processing and encapsulation add an estimated 250 kWh(e)/m^2. Assuming a 90% yield from wafers to completed modules gives a total energy content of 2170 kWh(e)/m^2 of cell area.

The time the cell would have to operate to pay back this energy invested in its depends on where it is deployed. With the equivalent of 5 h of peak sunshine per day on average and 12% encapsulated cell efficiency, the annual amount of energy generated would be 219 kWh(e)/m^2. Hence, the energy payback time would be just under 10 years. Less direct energy inputs such as the energy required to make the machines used in cell manufacture plus the energy required to market and install systems together with the energy content of associated power storage and conditioning equipment would increase this payback time further.

However, the important point is that the same processing steps that have precluded the widespread use of silicon cells in the past on economic grounds are also those which contribute most significantly to the energy content of the cells. Improved technology for producing silicon cells as described in Chapter 7 not only improves the economic viability but also reduces the energy required to manufacture a solar cell dramatically. With such technology, energy payback times are reduced from 10 years for the inefficient processing sequence outlined in the present chapter to a fraction of a year.

6.8 SUMMARY

The standard solar cell processing technology as developed for space cells and initially used for terrestrial cells has been described as have encapsulation schemes for packaging the cells in weatherproof modules. Several areas have been identified where improvements in the standard technology are desirable. Improved technologies for producing the cells are described in Chapter 7.

Cells produced for space and those produced in the past for terrestrial use with the standard processing described in this chapter have substantial amounts of energy invested in them compared to their generating capabilities. This is not the case for the improved technologies of Chapter 7.

EXERCISES

6.1. Draw a block diagram showing the major steps involved in converting quartzite to silicon solar cells.

6.2. For solar modules made up of 12% efficient silicon cells, estimate the difference in the operating temperature of cells under bright sunshine when the module is open-circuited and when it is delivering maximum power.

6.3. One solar cell has an open-circuit voltage of 0.55 V and a short-circuit current of 1.3 A. A second cell has values of 0.60 V and 1.0 A for these parameters. Assuming that both cells obey the ideal diode law, calculate the open-circuit voltage and short-circuit current of the combination when the cells are connected: (a) in parallel; (b) in series.

6.4. A hypothetical solar module consists of 40 identical solar cells, each giving an open circuit voltage of 0.60 V and a short-circuit current of 3 A under bright sunshine. The module is short-circuited under bright sunshine and one cell partially shaded. Assuming that the cells obey the ideal diode law and neglecting temperature effects, find the power dissipated in the shaded cell as a function of the fractional shading of the cell.

REFERENCES

[6.1] L. P. HUNT, "Total Energy Use in the Production of Silicon Solar Cells from the Raw Material to Finished Product," *Conference Record, 12th*

IEEE Photovoltaics Specialists Conference, Baton Rouge, 1976, pp. 347–352.

[6.2]   C. L. YAWS et al., "Polysilicon Production: Cost Analysis of Conventional Process," *Solid-State Technology*, January 1979, pp. 63–67.

[6.3]   H. YOO et al., "Analysis of ID Saw Slicing of Silicon for Low Cost Solar Cells," *Conference Record, 13th IEEE Photovoltaics Specialists Conference, Washington, D.C.*, 1978, pp. 147–151.

[6.4]   N. F. SHEPARD AND L. E. SANCHEZ, "Development of Shingle-Type Solar Cell Module," *Conference Record, 13th IEEE Photovoltaic Specialists Conference, Washington, D.C.*, 1978, pp. 160–164.

[6.5]   W. CARROL, E. CUDDIHY, AND M. SALAMA, "Material and Design Consideration of Encapsulants for Photovoltaic Arrays in Terrestrial Applications," *Conference Record, 12th Photovoltaic Specialists Conference, Baton Rouge*, 1976, pp. 332–339.

[6.6]   J. W. STULTZ AND L. C. WEN, *Thermal Performance, Testing and Analysis of Photovoltaic Modules in Natural Sunlight*, JPL Report No. 5101-31, July 1977.

[6.7]   E. ANAGNOSTOU AND A. F. FORESTIERI, "Endurance Testing of First Generation (Block 1) Commercial Solar Cell Modules," *Conference Record, 13th IEEE Photovoltaic Specialists Conference, Washington, D.C.*, 1978, pp. 843–846.

[6.8]   M. MACK, "Solar Power for Telecommunications," *Telecommunications Journal of Australia 29*, No. 1 (1979), 20–44.

[6.9]   C. GONZALEZ AND R. WEAVER, "Circuit Design Considerations for Photovoltaic Modules and Systems," *Conference Record, 14th IEEE Photovoltaics Specialists Conference, San Diego*, 1980, pp. 528–535.

Chapter

IMPROVED SILICON CELL TECHNOLOGY

7.1 INTRODUCTION

In Chapter 6, the standard technology used in the past to produce silicon solar cells was described. In converting quartzite, the source material, to encapsulated cells, several steps were seen to be costly and energy-intensive.

Activity worldwide is being directed at reducing these costs. The present chapter describes some of the most promising new silicon technology for each of the processing steps outlined in Chapter 6. Most of this technology is at an advanced stage of development. Some aspects are being evaluated at the pilot production stage, whereas others have already been incorporated into commercial products.

7.2 SOLAR-GRADE SILICON

It was seen in Chapter 6 that solar cells presently use the ultrapure silicon produced for the semiconductor industry. However, in transistors and integrated circuits, the emphasis is on silicon quality, and material costs are relatively unimportant. For solar cells, it is worthwhile considering a trade-off between performance and cost.

As mentioned in Section 3.4.4, impurities in solar cells generally introduce allowed levels into the forbidden gap and thereby act as recombination centers. It was seen in Section 5.4.2 that an increased density of such centers will decrease the cell efficiency. Figure 7.1 shows the experimental results for a range of different metallic impurities, when each is the only impurity present apart from the dopant (Ref. 7.1). Although some metallic impurities (Ta, Mo, Nb, Zr, W, Ti, and V) can reduce cell performance when present in extremely small concentrations, others can be present in concentrations in excess of $10^{15}/cm^3$ before becoming a problem. This is about 100 times higher than impurity levels in semiconductor-grade silicon (SeG-Si). It makes it likely that an alternative, less expensive process might produce a less pure *solar-grade* silicon (SoG-Si) which still gives cells of adequate performance. Several alternative processes appear capable of producing silicon of quality not appreciably lower than Se-G but at a fraction of the cost of conventional technology.

One of the more promising is that developed by the Union Carbide Corporation. It involves the preparation of silane (SiH_4) from

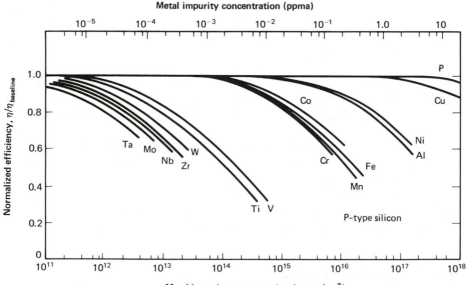

Figure 7.1. Effect of different secondary impurities on the performance of silicon solar cells (After Ref. 7.1, ©1978 IEEE.)

metallurgical-grade silicon and the subsequent deposition of silicon from silane (Ref. 7.2). Analysis of this process indicates that it is capable of producing silicon at one-fifth the cost of the present commercial process using only one-sixth of the processing energy. An alternative process at an advanced stage of development is that developed by Batelle Columbus Laboratories. It is based on the zinc reduction of silicon tetrachloride (Ref. 7.3). It is probable that improved processes such as these will replace the conventional Siemens process for producing silicon not only for the solar cell industry but also for the semiconductor electronics industry in general.

7.3 SILICON SHEET

7.3.1 Sheet Requirements

Having produced purified silicon, it is then necessary to convert it into the form of thin sheets of good crystallographic quality for use as solar cells. To obtain the full photovoltaic output from the material, it only has to be 100 μm or so thick. In the past, the approach used has been to form a large single-crystal ingot using the Czochralski (CZ) process and then slice thin wafers from this ingot. This approach is very inefficient in converting bulk silicon into large areas of solar cells. Not only is over half the silicon lost in sawing it into wafers, but cutting limitations result in thicker wafers than essential. Also, the wafers are circular, which means that they cannot be packed very densely when encapsulated into solar modules unless trimmed to a square or hexagonal shape.

7.3.2 Ingot Technologies

The Czochralski approach is an example of an *ingot technology*. The fundamental limitation of such a technology is that the ingots produced have to be sliced up into wafers with the disadvantages already mentioned.

The standard Czochralski process can be modified to operate on a semicontinuous basis with resultant cost savings (Ref. 7.4). The fact that the process produces cylindrical ingots[1] is also a disadvantage in solar cell applications.

A simpler approach to producing ingots, particularly those of square cross section, is to use a process similar to casting. This will

---

[1] With modifications, ingots with an approximately square cross section can be prepared by the Czochralski process (Ref. 7.5).

generally result in a polycrystalline ingot which is not ideal for solar cell applications. However, with careful control over the conditions under which molten silicon solidifies, silicon of large grain size can be formed. With a suitable "mold" material, such large-grained material can produce solar cells of good performance (Ref. 7.6).

With a suitable seed and a method of controlling the rate of solidification as in the heat-exchanger method (Ref. 7.7), it is possible to produce essentially single-crystal ingots of quite massive proportions with a "casting" approach. The performance of cells fabricated from such material has been comparable to that of cells made of good-quality Czochralski material. Studies indicate that this approach has a marked economic advantage over even the advanced Czochralski techniques (Ref. 7.8).

7.3.3 Ribbon Silicon

The limitations of the ingot approach can be avoided if the silicon can be formed directly into sheets or ribbons. Several techniques have been developed for achieving this.

The first to be developed to the stage of producing commercial cells is the *edge-defined film-fed growth* (EFG) method, as illustrated in Fig. 7.2. This technique bears some relationship to the Czochralski process except that the shape of the crystal pulled is defined by a graphite "die." As a consequence, it is possible to obtain crystals in the form of thin ribbons directly from the melt (Ref. 7.9). Very high production rates of silicon can be obtained by pulling several ribbons simultaneously from the same melt (Ref. 7.10).

The major problem with this technique relates to the quality of the material produced. The process produces material of relatively poor crystallographic quality compared to the Czochralski process. Owing to the nature of the growth process, impurities introduced into the molten silicon from the die, crucible, and surrounding regions of the growth furnace are incorporated into the growing ribbon. (In other crystal growth processes, most of these are preferentially rejected from the growing crystal to the molten silicon.) In addition, the molten silicon tends to react with the graphite die. This can result in silicon carbide precipitates in the ribbon which can disrupt its growth and degrade the properties of subsequently fabricated cells.

The dendritic web method illustrated in Fig. 7.3 overcomes some of these disadvantages. By controlling temperature gradients, it is possible to encourage the growth of parallel dendrites into the molten material. As these are pulled out of the melt, a film of molten silicon is trapped between them (Ref. 7.11). This subsequently solid-

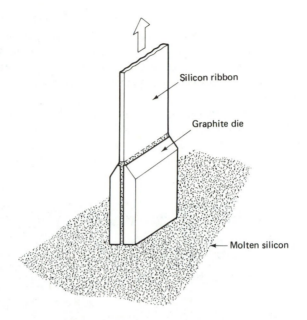

Figure 7.2.   Essential features of the edge-defined film-fed growth method for producing ribbon crystals of silicon. The molten silicon moves up the interior of the graphite die by capillary action. The ribbon shape is defined by the shape of the top of the die.

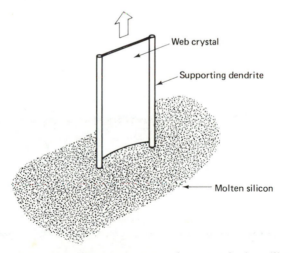

Figure 7.3.   Dendritic web approach to producing silicon ribbon.   No die is required.   The shape of the ribbon is controlled by thermal gradients in the molten silicon.   The dendrites down the edge of the ribbon solidify first.   As these are drawn out of the melt, a thin layer of initially molten silicon is trapped between them.

ifies to give a thin ribbon of silicon bounded by the thicker dendrites. These can be removed and recycled after cell fabrication on the ribbon section. No die is required and material properties are nearly as good as these obtained from the Czochralski process. The relatively low production rate of silicon crystal material by this process has been its major disadvantage.

Relatively small samples of silicon ribbon produced by both the EFG and dendritic web processes are compared in Fig. 7.4. Note the corrugated surface finish typical of material produced by the EFG process and the mirror finish characteristic of dendritic web material.

Figure 7.4.   Comparison of the physical appearance of EFG and dendritic web silicon ribbon crystals. The dendrites are removed from the latter after cell processing. (Ribbon courtesy of Japan Solar Energy Corporation and Westinghouse Research Laboratories.)

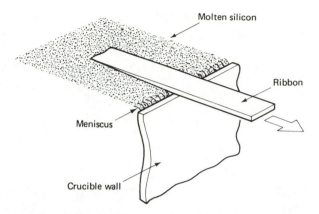

Figure 7.5. Horizontal or low-angle growth method for producing silicon ribbon. Ribbons can be produced at very high rates, but dimensional control has been a problem in the past.

Several other ribbon growth processes are at a less advanced stage of development. One distinguished by very high rates of ribbon production is the *horizontal* or *low-angle growth method* of Fig. 7.5. The major developmental problems with this technique have related to control of the ribbon dimensions (Ref. 7.12). Casting of silicon directly into sheet or ribbon form has also been explored (Refs. 7.6 and 7.13).

7.4 CELL FABRICATION AND INTERCONNECTION

The disadvantage of the standard technique for fabricating cells on silicon wafers as described in Chapter 6 is that it is a "batch"-orientated technique. As such, it is capable of only relatively low throughput and also is extravagant in the use of materials, particularly metal for forming the cell contacts. Several approaches are now being used in commercial devices which overcome these limitations and, as a bonus, result in cells of higher performance.

One major development has been the use of silicon wafers with *textured surfaces*. Using selective etching, it is possible to form small pyramids on the surface of silicon wafers (Ref. 7.14). The appearance of these pyramids greatly magnified under the scanning electron microscope is shown in Fig. 7.6. Light reflected from the side of one of these pyramids is reflected downward, getting a second chance of being coupled into the cell. With the use of an antireflection coating, reflection losses can be reduced to less than a few percent. As shown

127

Figure 7.6.   Appearance of a textured silicon surface under a scanning electron microscope.   The peaks are typically 10 μm high and are the tops of square-based pyramids.   The sides of these pyramids are intersecting (111) planes within the crystal structure of the silicon.

in Chapter 8, this technique can give acceptable performance even without antireflection coatings.

In the previous standard process, the *p-n* junction was formed by diffusing impurities on a batch basis.   A more cost-effective approach is to spray layers containing the required dopant onto the wafer surface and to diffuse the impurity in using a continuous-belt process (Ref. 7.15).   An alternative approach is to use a technique known as *ion implantation* (Ref. 7.16).   Ions of the dopant are accelerated to high velocities and directed at the wafer surface.   These are embedded in the silicon wafer near the surface they impact.   A subsequent annealing step removes the damage inflicted on the silicon lattice by this process and also "activates" the dopants in an electronic sense.   Pulsing by an electron beam or laser is an energy-efficient annealing approach.   A very tidy process for the high-volume production of solar cells based on ion implantation is indicated in Fig. 7.7.

Further improvement in solar cell performance has been obtained by paying attention to obtaining a low-surface-recombination-velocity contact to the rear of the cell.   As indicated in Chapter 5, this improves the open-circuit voltage of the cell and it also marginally increases the current output.   The use of what is somewhat

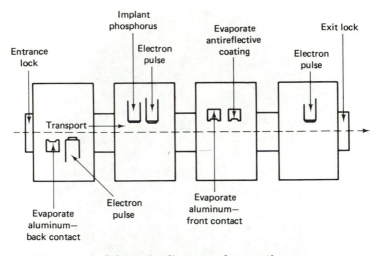

Figure 7.7. Schematic diagram of a continuous vacuum process for producing solar cells based on ion-implanted junctions, evaporated metal contacts, and pulsed electron beam annealing. (After Ref. 7.16, © 1976 IEEE.)

inappropriately called a *back surface field* (BSF) is one technique for achieving low effective recombination velocities at the rear of the cell. As indicated in Fig. 7.8, a heavily doped region is included right at the back contact to the cell. From theoretical considerations, the interface between this region and the more lightly doped bulk regions can be shown to act as a low-recombination-velocity surface. Not only does this technique improve voltage and current output as mentioned, but it also makes it easier to make a low-resistance contact to the silicon at the rear of the cell. In practice, the most effective technique for producing this "back surface field" has been to screen print an alu-

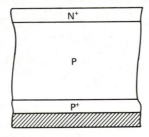

Figure 7.8. Sketch of the N^+PP^+ solar cell. The heavily doped P^+ region at the rear of the cell blocks minority carrier flows, making the PP^+ junction plane an effective low-recombination-velocity surface.

129

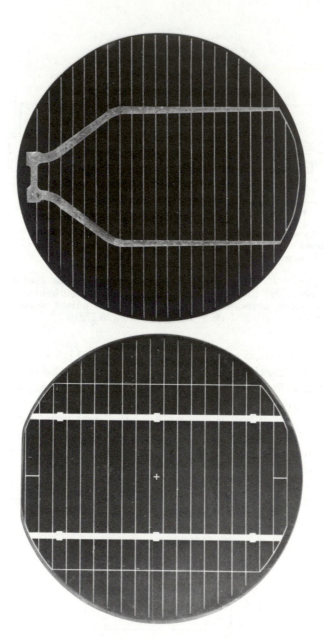

Figure 7.9.   Solar cells of 10 cm in diameter with solder-dipped (upper) and screen-printed (lower) top-layer metallization.

minum-based paste onto the rear of the cell and alloy the aluminum into the silicon in a subsequent firing step (Ref. 7.15).

Metallization was one of the weak links in the standard solar cell processing sequence of Chapter 6. Two lower-cost processes used in commercial cells are screen printing and electroplating. Both of these prevent wastage of metal and eliminate the requirement for expensive vacuum equipment. In the former, a paste containing the metal is printed onto the silicon wafer through a mask and subsequently fired to remove "binders" in the paste and to lower the metal resistivity. Silver pastes were the first to be used commercially although nickel, aluminum, and copper pastes may be lower-cost alternatives. In the plating approach, a pattern is etched through an insulating layer on the cell surface and the desired metal layer is plated through this. More than one type of metal layer generally is plated sequentially, since few metals by themselves possess the properties of good adherence to silicon, corrosion resistance, low resistivity, and low cost. A final step in this approach may be "solder dipping" to cover the plated metal by a layer of solder for corrosion protection and lower series resistance. Figure 7.9 compares the physical appearance of cells with solder-dipped and screen-printed top-layer metallization.

Spraying appears to be the most cost-effective method for applying the antireflection coating, although with textured surfaces this layer may not be required. Automated machinery has been developed for connecting the cells together and for their encapsulation into solar modules (Ref. 7.17).

7.5 ANALYSIS OF CANDIDATE FACTORIES

Several alternative approaches to producing silicon cells have been outlined in previous sections. To provide common ground for the comparison of economic aspects of these approaches, a costing technique known as SAMICS (Solar Array Manufacturing Industry Costing Standards) has been developed (Ref. 7.18). Using a computer program based on this technique, some of the best documented and validated costing has been done of the processing sequences involved in fabricating solar modules.

Large-capacity factories have been designed based on different combinations of the technologies described in this chapter and the cost of the resulting solar modules compared using the SAMICS methodology. The results have been that several combinations of these technologies appear capable of producing solar modules at costs which would make them competitive energy generators in some of the large-scale applications described in Chapter 14. The EFG and dendritic web ribbon approaches as well as HEM (heat-exchanger method) with subsequent wafer slicing all appear viable silicon sheet technologies, with the advanced Czochralski approaches lagging somewhat behind (Ref. 7.8).

As an example of an automated factory capable of producing low-cost silicon modules, the characteristics of one of the first to be analyzed by the SAMICS approach will be described (Ref. 7.19). The product resulting from this factory is a 1.2 × 1.2 m module made up of 192 cells fabricated on EFG ribbon silicon cut to a size of 10 × 7.5 cm as shown in Fig. 7.10. The module efficiency is 11.4%.

The processing sequence used to produce the module is shown in Fig. 7.11. The Union Carbide process is used to refine metallurgical-grade silicon. Silicon sheet is prepared from this material by the EFG process and cut to size. The major stages in fabricating cells on this material are the application of a back surface field, texture etching, ion implantation of the junction followed by pulse annealing, and the screen printing of contacts. The cells are then encapsulated and tested.

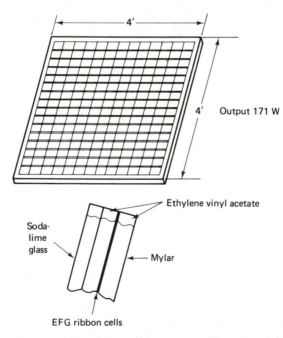

Figure 7.10.   Solar cell module produced by the candidate factory described in the text. (After Ref. 7.19.)

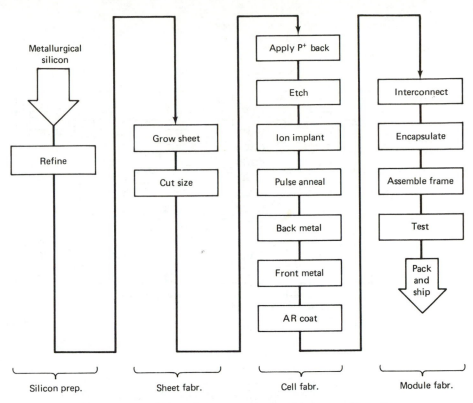

Figure 7.11. Processing sequence used to produce the module of Figure 7.10. (After Ref. 7.19.)

Production rate (W/yr)	250,000,000
Labor force (all shifts)	1,152 direct
	529 indirect

Factory area (ft^2)	
Silicon refinement	40,800
Sheet growth	19,550
Cell fabrication	31,200
Module fabrication	10,200
Warehouse	9,779
Misc. (aisles, shops, cafeteria, etc.)	20,901

Capital equipment ($)	
Silicon refinement	19,400,000 *
Sheet growth	14,820,000
Cell and module fabrication	8,219,000

ADM

Silicon refinement 40,800 ft^2

Sheet, cell, and module fabrication 91,630 ft^2

*Union Carbide
Energy payback time sheet, cell, and module 0.179 year

Figure 7.12. Calculated space, labor, and capital requirements (1975 $US) to produce cells in a volume of 250 MW$_p$/yr. The calculated selling price of the module (F.O.B. factory) was $0.46 per W$_p$ in the same monetary units. (After Ref. 7.19.)

133

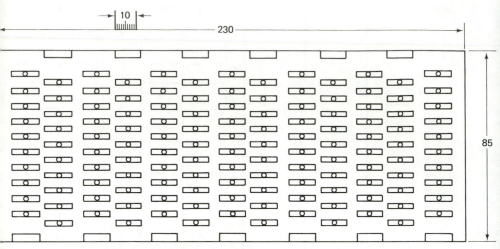

130 pairs of 5-ribbon machines with melt replenisher between
10 pairs with maintenance bench per column
One operator per pair

(a)

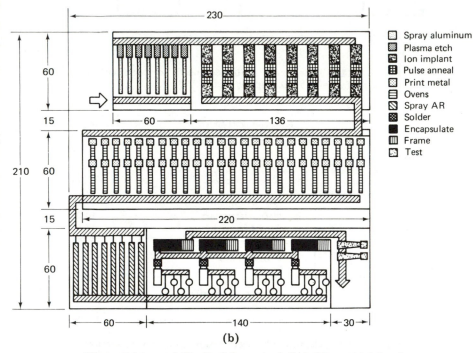

Spray aluminum
Plasma etch
Ion implant
Pulse anneal
Print metal
Ovens
Spray AR
Solder
Encapsulate
Frame
Test

(b)

Figure 7.13. (a) Physical layout of the ribbon crystal pullers for the candidate factory. Each puller produces five ribbons simultaneously using the EFG approach (dimensions are in feet). (b) Physical layout of cell processing, encapsulation, and testing sections of the plant. (After Ref. 7.19.)

To achieve the low costs mentioned, production volume has to be an estimated 250 MW_p/yr. The labor requirements and area requirements for different steps in the module fabrication are shown in Fig. 7.12. The operating time of the modules to pay back the processing energy invested in them for this processing sequence is a very low 65 days.

To give an idea of the size and physical layout of different sections of the candidate 250 MW_p/yr factory, Fig. 7.13(a) shows a possible layout for the sheet-preparation section of the factory. Figure 7.13(b) shows the sections dealing with cell fabrication, encapsulation, and testing.

7.6 SUMMARY

Several advanced technologies have been developed for producing low-cost silicon solar cell modules. These eliminate the deficiencies in the standard processing sequence described in Chapter 6.

Direct growth of silicon ribbons eliminates wafer slicing, which is the weak link in any ingot technology. The main requirements on techniques for fabricating cells on silicon wafers are that they be capable of a high degree of automation and that they do not consume excessive materials. Several processing sequences satisfy these criteria. Economic analysis of candidate factories based on advanced processing sequences show that it should be possible to produce silicon solar cell modules at prices that would make them competitive with conventional sources of electricity, at least in *some* large-scale applications. A "thin-film" technology as described in Chapter 9 places less severe demands on the quantity of photovoltaic material required. For this reason, approaches other than those based on crystalline silicon, as described in this chapter, will prove even less expensive in the long term.

EXERCISES

7.1. The volume density of silicon atoms in a silicon crystal is about 5×10^{28} m^{-3}. According to Fig. 7.1, how many parts per billion of the following impurities can be tolerated in standard solar cells without degrading performance by more than 10%? (a) Mo; (b) Ti; (c) Cu.

7.2. (a) The best performance in the near term expected to be able to be maintained in a production setting with advanced silicon ingot slicing techniques is to produce wafers 250 μm thick with a cutting or kerf loss of 150 μm. With such a technology, what is the maximum area of silicon that can be obtained from each kilogram of silicon starting material, assuming that the material lost in cutting is not reclaimed?

(b) What is the corresponding figure for a ribbon technology that produces ribbons 100 μm thick?

(c) If the material produced as above results in cells of 16% and 12% energy-conversion efficiency, respectively, under bright sunlight (1 kW/m^2), find a value for the maximum power-generating capacity of each of the foregoing technologies expressed as peak watts/kilograms.

(d) If pure silicon were to be produced at the present volume of metallurgical-grade silicon ($\sim$1000 million kilograms annually), what would be the maximum possible generating capacity from one year's production for solar cells made by each of the foregoing approaches?

7.3. Calculate the maximum packing density (cell area to module area) that can be obtained for circular cells packed into rectangular modules.

REFERENCES

[7.1] J. R. Davis et al., "Characterization of the Effects of Metallic Impurities on Silicon Solar Cell Performance," *Conference Record, 13th IEEE Photovoltaic Specialists Conference, Washington, D.C.*, 1978, pp. 490–496.

[7.2] C. L. Yaws et al., "New Technologies for Solar Energy Silicon: Cost Analysis of UCC Silane Process," *Solar Energy 22* (1979), 547–553.

[7.3] C. L. Yaws et al., "New Technologies for Solar Energy Silicon: Cost Analysis of BCL Process," *Solar Energy 24* (1980), 359–365.

[7.4] G. F. Fiegl and A. C. Bonora, "Low Cost Monocrystalline Silicon Sheet Fabrication for Solar Cells by Advanced Ingot Technology," *Conference Record, 14th IEEE Photovoltaic Specialists Conference, San Diego*, 1980, pp. 303–308; A. H. Kachare et al., "Performance of Silicon Solar Cells Fabricated from Multiple Czochralski Ingots Grown by Using a Single Crucible," *Conference Record, 14th IEEE Photovoltaic Specialists Conference, San Diego*, 1980, pp. 327–331.

[7.5] J. C. Posa, "Motorola Pulls Square Ingots," *Electronics*, October 11, 1979, p. 43.

[7.6] H. Fischer and Pschunder, "Low Cost Solar Cells Based on Large Area Unconventional Silicon," *Conference Record, 12th IEEE Photovoltaic Specialists Conference, Baton Rouge*, 1976, pp. 86–92; J. Lindmayer and Z. C. Putney, "Semicrystalline versus Single Crystal Silicon," *Conference Record, 14th IEEE Photovoltaic Specialists Conference, San Diego*, 1980, pp. 208–213.

[7.7] C. P. Khattak and F. Schmid, "Low-Cost Conversion of Polycrystalline Silicon into Sheet by HEM and Fast," *Conference Record, 14th IEEE Photovoltaic Specialists Conference, San Diego*, 1980, pp. 484–487.

[7.8] R. W. ASTER, "PV Module Cost Analysis," in *LSA Project Progress Report 13 for Period April 1979 to August 1979*, DOE/JPL-1012-29, pp. 3-385 to 3-395.

[7.9] K. V. RAVI, "The Growth of EFG Silicon Ribbons," *Journal of Crystal Growth 39* (1977), 1-16.

[7.10] J. P. KALEJS et al., "Progress in the Growth of Wide Silicon Ribbons by the EFG Technique at High Speed Using Multiple Growth Stations," *Conference Record, 14th IEEE Photovoltaic Specialists Conference, San Diego*, 1980, pp. 13-18.

[7.11] R. G. SEIDENSTICKER, "Dendritic Web Silicon for Solar Cell Application," *Journal of Crystal Growth 39* (1977), 17-22.

[7.12] T. KOYANAGI, "Sunshine Project R and D Underway in Japan," *Conference Record, 12th IEEE Photovoltaic Specialists Conference, Baton Rouge*, 1976, pp. 627-633; D. N. JEWETT AND H. E. BATES, "Low Angle Crystal Growth of Silicon Ribbon," *Conference Record, 14th IEEE Photovoltaic Specialists Conference, San Diego*, 1980, pp. 1404-1405.

[7.13] D. J. ROWCLIFFE AND R. W. BARTLETT, "Vacuum Die Casting of Si Sheet," in *LSA Progress Report 13 for Period April 1979 to August 1979*, DOE/JPL-1012-39, pp. 3-152 to 3-154.

[7.14] S. R. CHITRE, "A High Volume Cost Efficient Production Macrostructuring Process," *Conference Record, 13th IEEE Photovoltaic Specialists Conference, Washington, D.C.*, 1978, pp. 152-154.

[7.15] N. MARDESICH et al., "A Low-Cost Photovoltaic Cell Process Based on Thick Film Techniques," *Conference Record, 14th IEEE Photovoltaic Specialists Conference, San Diego*, 1980, pp. 943-947.

[7.16] A. KIRKPATRICK et al., "Silicon Solar Cells by Ion Implantation and Pulsed Energy Processing," *Conference Record, 12th IEEE Photovoltaic Specialists Conference, Baton Rouge*, 1976, pp. 299-302; also "Low-Cost Ion Implantation and Annealing Technology for Solar Cells," *Conference Record, 14th IEEE Photovoltaic Specialists Conference, San Diego*, 1980, pp. 820-824.

[7.17] H. SOMBERG, *Automated Solar Panel Assembly Line*, report prepared for Jet Propulsion Laboratory, Report No. DOE/JPL/955278-1, April 1979, and subsequent reports.

[7.18] R. G. CHAMBERLAIN, "Product Pricing in the Solar Array Manufacturing Industry: An Executive Summary of SAMICS," *Conference Record, 13th IEEE Photovoltaic Specialists Conference, Washington, D.C.*, 1978, pp. 904-907.

[7.19] J. V. GOLDSMITH AND D. B. BICKLER, *LSA Project—Technology Development Update*, report to U.S. Department of Energy by Jet Propulsion Laboratory, DOE/JPL-1012-7, August 1978 (JPL Pub #79-26); also D. B. BICKLER, "A Preliminary 'Test Case' Manufacturing Sequence for 50^c/Watt Solar Photovoltaic Modules in 1986," *Proceedings of Second E. C. Photovoltaic Solar Energy Conference* (Utrecht: D. Reidel Publishing Co., 1979), pp. 835-842.

Chapter

DESIGN
OF SILICON SOLAR CELLS

8.1 INTRODUCTION

In previous chapters, standard and improved technologies for producing silicon solar cells have been described. In the present chapter, considerations relevant to the detailed design of silicon cells will be discussed. Answers will be found to questions such as: What is the optimum level of dopants on either side of the junction? Where is the best
location for the junction? What is the best shape for the top contact
to the cell? How can optical losses from the cell be minimized?

Although answers to these questions will be provided specifically for silicon cells, parallel considerations will apply to cells made
from other materials discussed in Chapter 10.

8.2 MAJOR CONSIDERATIONS

8.2.1 Collection Probability
of Generated Carriers

A spatially dependent parameter, the *collection probability*,
can be defined as the *probability a light-generated minority carrier*

138

has of contributing to the short-circuit current of a solar cell. This is a function of the position the carrier is generated within the cell. This parameter will be seen to be critical in determining the physical design of solar cells.

To find the collection probability, the artificial situation shown in Fig. 8.1(a) will be analyzed. Generation of electron–hole pairs by light will be assumed to occur only at points lying on a single plane throughout the cell. For the case where symmetry allows a "one-dimensional" analysis, the generation rate as a function of distance through the cell will be an impulse function, as indicated in Fig. 8.1(b).

The aim of the analysis will be to find the proportion of electrons generated at the point x_1 which contribute to the corresponding current flowing in the cell under short-circuit. During the analysis, it will be seen that no nonlinearities are involved and, by the superposition principle, the results can be applied to cases where the form of the generation rate corresponds more closely to practical situations. The analysis closely parallels that of Section 4.6.

In region 1 of Fig. 8.1(b), the generation rate is zero everywhere except at the single point x_1 right at the edge of the region. The differential equation that the excess minority carriers, Δn, must satisfy is therefore similar to Eq. (4.25):

$$\frac{d^2 \Delta n}{dx^2} = \frac{\Delta n}{L_e^2} \tag{8.1}$$

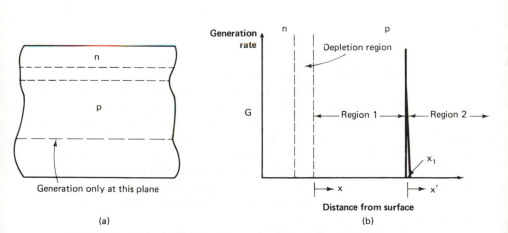

Figure 8.1. Idealized carrier-generation conditions used to calculate collection probabilities. An expression for the fraction of carriers generated at point x_1 which contribute to the cell short-circuit current is found in the text.

where L_e is the diffusion length. The general solution is, as before,

$$\Delta n = A e^{x/L_e} + B e^{-x/L_e} \tag{8.2}$$

Boundary conditions are applied to assign values to the constants A and B. Under short-circuit conditions, the excess electron concentration at $x = 0$ is zero, since its value is determined by the voltage across the junction. Hence, $A = -B$ and

$$\Delta n = A(e^{x/L_e} - e^{-x/L_e}) = 2A \sinh\left(\frac{x}{L_e}\right) \tag{8.3}$$

Similarly, in region 2 of Fig. 8.1(b),

$$\Delta n = C e^{x'/L_e} + D e^{-x'/L_e}$$

where the origin of the x' direction is at the point x_1. In this case, the excess concentration must be finite as x' becomes large. Hence, $C = 0$ and

$$\Delta n = D e^{-x'/L_e} \tag{8.4}$$

At the position x_1, the two solutions given by Eqs. (8.3) and (8.4) must match up since the electron concentration is continuous. Hence,

$$D = 2A \sinh\left(\frac{x_1}{L_e}\right) \tag{8.5}$$

On the n-type side of the device, the generation rate is zero throughout since the only generation occurs at x_1. Also, the excess hole concentration Δp is zero at the edge of the depletion region when the cell is short-circuited. It follows that Δp is zero throughout this region. The resulting distributions of electrons and holes are shown in Fig. 8.2(a). Since minority carriers in uniformly doped quasi-neutral regions flow predominantly by diffusion (Section 4.5), minority-carrier current flows can be readily calculated by differentiating the foregoing distributions. The results are illustrated in Fig. 8.2(b).

The discontinuity in the electron current density at the point x_1 arises from the generation of carriers at this point. The size of the discontinuity is just the electronic charge multiplied by the rate of generation at this point. Assuming little change in currents across

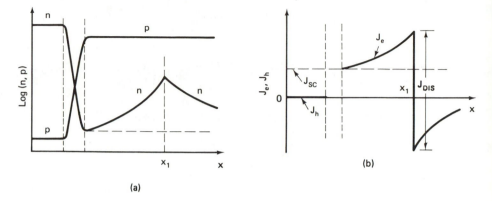

Figure 8.2. (a) Carrier distributions through the solar cell for the generation conditions of Fig. 8.1. (b) Corresponding minority carrier current distributions.

the depletion region (as previously discussed in Sections 4.6.2 and 5.4.3), the total current density in the device is equal to the electron current density at $x = 0$. The probability of collection (f_c) is just the ratio of the rate of flow of carriers in the external circuit to the rate of generation. Hence,

$$f_c = \frac{J_{SC}}{J_{DIS}} \tag{8.6}$$

However, in the p-type region,

$$J_e = qD_e \frac{dn}{dx} \tag{8.7}$$

Hence, in region 1,

$$J_e = \frac{2qD_e A}{L_e} \cosh\left(\frac{x}{L_e}\right) \tag{8.8}$$

which gives, at $x = 0$,

$$J_{SC} = \frac{2qD_e A}{L_e} \tag{8.9}$$

J_{DIS} can be found by evaluating the two expressions for current flow on either side of the discontinuity (J_{e-} and J_{e+}):

141

$$J_{e-} = \frac{2qD_eA}{L_e} \cosh\left(\frac{x_1}{L_e}\right) \tag{8.10}$$

$$J_{e+} = \frac{-qD_eD}{L_e} = \frac{-2qD_eA}{L_e} \sinh\left(\frac{x_1}{L_e}\right) \tag{8.11}$$

The latter equation follows from Eqs. (8.5) and (8.7). Hence,

$$J_{\text{DIS}} = J_{e-} - J_{e+}$$

$$= \frac{2qD_eA}{L_e} e^{x_1/L_e} \tag{8.12}$$

It follows that

$$\boxed{f_c = e^{-x_1/L_e}} \tag{8.13}$$

The collection probability decreases exponentially with increasing distance of the point of generation away from the edge of the junction-depletion region. The characteristic decay length is just the minority carrier diffusion length. Since the foregoing analysis is linear, this conclusion is valid regardless of the distribution of the generation rate of carriers throughout the device.

A sketch of the probability of collection as a function of distance through a solar cell is shown in Fig. 8.3. As postulated earlier (Section 4.7), the depletion region and the region of the cell lying

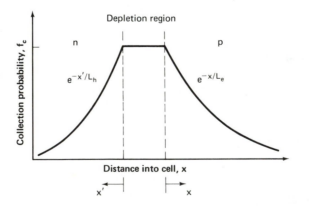

Figure 8.3. Calculated collection probability of generated minority carriers as a function of the point of generation in the solar cell.

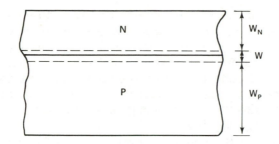

Figure 8.4.   Critical dimensions in a cell of finite extent.

within a minority-carrier diffusion length of it are the regions where generated carriers have the highest chance of contributing to current flows (i.e., of being collected).   A minority carrier generated outside these regions has an excellent chance of recombining before reaching the junction and subsequently the device terminals.

The foregoing analysis has assumed implicitly that both the n- and p-type regions extend much further than a diffusion length from the junction.   For actual cells of finite extent as shown in Fig. 8.4, the collection probability will be modified.   For example, if the surface on the p-type side of the device has a high recombination velocity, the expression corresponding to Eq. (8.13) is

$$f_c = \frac{\sinh\,[(W_P - x)/L_e]}{\sinh\,(W_P/L_e)} \qquad (8.14)$$

If it has a low surface-recombination velocity, then

$$f_c = \frac{\cosh\,[(W_P - x)/L_e]}{\cosh\,(W_P/L_e)} \qquad (8.15)$$

Both these approach the expression of Eq. (8.13) for $W_P \gg L_e$.

8.2.2   Junction Depth

Exposed surfaces such as between grid lines at the top of the solar cell are generally high-recombination-velocity surfaces.   Interfaces between ohmic metal contacts and semiconductor are also generally regions of high recombination velocity.   It has already been shown in Section 7.4 that one way of creating an effective low-recombination-velocity interface was by using a back surface field (i.e., by forming a junction between highly doped and lightly doped material of the same dopant type).

143

In view of the above and Eqs. (8.14) and (8.15), the collection probability of generated carriers as a function of distance from the surface of a *p-n* junction solar cell has the form shown in Fig. 8.5(a). Two features are important. A back surface field improves the probability of collection of carriers generated near the back contact and hence improves the short-circuit current of the cell. The probability of collection of carriers generated near the top surface of the solar cell is generally low.

Considering the actual generation rate of carriers in a semiconductor when illuminated by sunlight, the highest rate of generation will occur right at the semiconductor surface. For monochromatic light, the generation rate is given by

$$G = (1 - R)\, \alpha N e^{-\alpha x} \qquad (8.16)$$

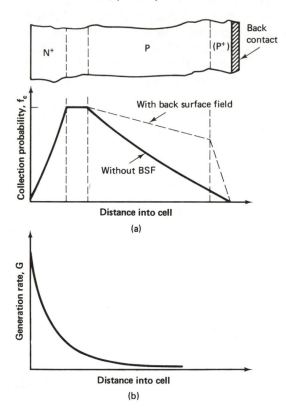

Figure 8.5. (a) Collection probability for a cell of finite extent with and without a back surface field. (b) Schematic diagram showing the generation rate of electron–hole pairs as a function of distance into the cell when illuminated by sunlight.

where x is the distance below the surface, α the absorption coefficient, N the incident photon flux, and R the fraction reflected. Under sunlight, the generation rate becomes

$$G(x) = \int_0^{\lambda_{max}} [1 - R(\lambda)] \, \alpha(\lambda) \, N'(\lambda) \, e^{-\alpha(\lambda)x} \, d\lambda \qquad (8.17)$$

where $N'(\lambda)$ is the incident flux per unit wavelength. The approximate form of this weighted exponential is shown in Fig. 8.5(b). The generation rate has a very strong peak near the surface precisely where the probability of collection is low. *This deficiency obviously is minimized if the junction is as close to the surface as possible.*

8.2.3 Lateral Resistance of Top Layer

Current flows in the bulk of the device are generally perpendicular to the surface of the cell as indicated in Fig. 8.6(a). To extract current via a contact that partially covers the top of the cell, current must flow laterally through the top layer of cell material. For a uniformly doped n-type layer, the resistivity of this layer is given by (Section 2.14)

$$\rho = \frac{1}{q\mu_e N_D} \qquad (8.18)$$

A quantity more appropriate for describing the lateral resistance of this layer is the "sheet resistivity" (ρ_s), which is the resistivity divided by the layer thickness, t:

$$\rho_s = \frac{1}{q\mu_e N_D t} \qquad (8.19)$$

For nonuniformly doped layers, the product $\mu_e N_D t$ is replaced by the integral $\int_0^t \mu_e(x) \, N_D(x) \, dx$. Sheet resistivities have the dimensions of ohms but are normally expressed as ohms/square ($\Omega/\square$).

The sheet resistivity determines the minimum spacing required between the grid lines of the top contact. Referring to Fig. 8.6(b), the resistive power loss due to lateral current flow is readily calculable. The incremental power loss in the section dy is given by

$$dP = I^2 \, dR \qquad (8.20)$$

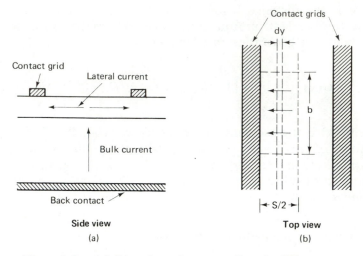

Figure 8.6. (a) Direction of current flow in different regions of a *p-n* junction solar cell. (b) Calculating power loss due to the lateral resistance of the top layer.

where dR is just $\rho_s dy/b$, and I, the lateral current flow, is zero at the midpoint between grating lines and increases linearly to its maximum at the grating line under uniform illumination. Hence,

$$I = Jby \qquad (8.21)$$

where J is the current density in the device. The total power loss is the integral of the incremental losses:

$$P_{\text{loss}} = \int I^2 dR = \int_0^{S/2} \frac{J^2 b^2 y^2 \rho_s dy}{b} = \frac{J^2 b \rho_s S^3}{24} \qquad (8.22)$$

At the maximum power point, the generated power from the region under consideration is $V_{\text{mp}} J_{\text{mp}} bS/2$. Hence, the fractional power loss at this point is

$$\boxed{p = \frac{P_{\text{loss}}}{P_{\text{mp}}} = \frac{\rho_s S^2 J_{\text{mp}}}{12 V_{\text{mp}}}} \qquad (8.23)$$

This allows the minimum spacing S to be found for a given set of cell parameters. For example, for a typical commercial silicon cell, $\rho_s = 40\ \Omega/\square$, $J_{\text{mp}} = 30\ \text{mA/cm}^2$, and $V_{\text{mp}} = 450\ \text{mV}$. For less

146

than 4% power loss due to lateral resistance effects,

$$S^2 < \frac{12 \, p V_{\text{mp}}}{\rho_s J_{\text{mp}}}$$

That is,

$$S < \left(\frac{12 \times 0.04 \times 0.45}{40 \times 0.03} \right)^{1/2} \text{cm} < 4 \text{ mm} \qquad (8.24)$$

This agrees with the grid spacings found on commercial silicon cells. Those with any larger spacing have a smaller sheet resistivity, and those with smaller spacing have a larger sheet resistivity. In practice, *the major factor determining the sheet resistivity is the depth of the junction* in Eq. (8.19). In fact, the resolution limits of the technology used to produce the grid pattern effectively sets the lower limit to the depth of the junction beneath the cell surface. To achieve minimum sheet resistivity of this layer, it is doped as high as practical.

8.3   DOPING OF THE SUBSTRATE

The substrate is uniformly doped during its preparation from molten material. The level of doping used is determined by the criteria which will now be discussed.

To get maximum I_{SC} once the junction depth is specified, the key parameter is the diffusion length in the substrate material. This is determined primarily by the minority-carrier lifetime in this region ($L_e = \sqrt{D_e \tau_e}$). In Section 3.4, three different recombination mechanisms were identified which determine the value of this quantity. For all of these, the general trend is a decrease in lifetime with increasing doping level. This is indicated in Fig. 8.7, which shows this dependency and the relative values of the lifetimes as caused by the different processes. For recombination by traps, the ideal expression for the variation of lifetime with doping follows from Eq. (3.22). It has the form

$$\tau_{nT} = \tau_{n0} \left(1 + \frac{m_1}{N_A} \right) \qquad (8.25)$$

where m_1 is approximately equal to the larger of the parameters p_1 and $\tau_{p0} n_1 / \tau_{n0}$. Its value is determined by the energy and perhaps the

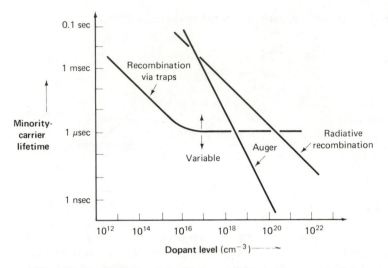

Figure 8.7. Relative magnitudes and dependence upon dopant density of minority-carrier lifetimes as determined by three competing processes for silicon.

capture cross sections of the dominant trap. For Auger recombination, the approximate expression at high doping levels is (Section 3.4.3)

$$\tau_{nA} = \frac{1}{DN_A^2} \tag{8.26}$$

while for radiative recombination (Section 3.4.2),

$$\tau_{nR} = \frac{1}{2BN_A} \tag{8.27}$$

The net recombination rate is given by

$$\frac{1}{\tau_n} = \frac{1}{\tau_{nT}} + \frac{1}{\tau_{nA}} + \frac{1}{\tau_{nR}} \tag{8.28}$$

As a consequence of the discussion above, it is concluded that increasing N_A will tend to decrease I_{SC}. Regarding the open-circuit voltage, the simple form of the expression for the saturation current density of a diode is given by Eq. (4.37)

$$I_0 = qA \left(\frac{D_e n_i^2}{L_e N_A} + \frac{D_h n_i^2}{L_h N_D} \right) \tag{8.29}$$

148

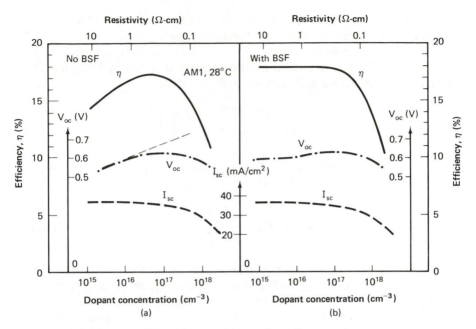

Figure 8.8. Dependence of key solar cell parameters upon the density of p-type dopants as obtained by high-performance experimental devices:

(a) Without a back surface field.
(b) With a back surface field.

The smaller I_0, the larger V_{oc}. Hence, it would appear appropriate to make N_A and N_D as large as possible to obtain maximum V_{oc}. For a p-type substrate, the doping in the n-type diffused region (N_D) is made as large as possible to reduce the sheet resistivity. Hence, the second component in Eq. (8.29) is reasonably small, as discussed further in Section 8.6. It is concluded that V_{oc} will tend to increase with increasing N_A.

Because of the opposite dependencies of I_{SC} and V_{oc} upon N_A, it follows that there will be an optimum substrate doping level for maximum energy-conversion efficiency. This is in agreement with experimental results as indicated in Fig. 8.8(a). The results shown are indicative of experimental devices of good performance on substrates of different doping levels.

8.4 BACK SURFACE FIELDS

It has been mentioned that a highly doped region near the back contact will increase the short-circuit current and the open-circuit volt-

149

age. It improves the former by virtue of improved collection efficiency in the vicinity of the back contact, as indicated in Fig. 8.5. It improves the latter by virtue of a reduced saturation current (Section 4.9). The contribution to the saturation current from a p-type substrate with such a BSF takes the form

$$I_{0p} = \frac{qD_e n_i^2}{L_e N_A} \tanh \left(\frac{W_P}{L_e} \right) \qquad (8.30)$$

Once the thickness of the p-type region is much less than a diffusion length ($W_P \ll L_e$), this equation reduces to

$$I_{0p} = \frac{qn_i^2 W_P}{\tau_e N_A} \qquad (8.31)$$

With the increase of τ_e with decreasing N_A indicated in Fig. 8.7, this means that V_{oc} will not be dependent on resistivity for the higher resistivities. This differs from the case without a back surface field. Maximum efficiencies will occur at lower doping densities, provided that the series resistance of the substrate is not a problem.[1] This is indicated in Fig. 8.8(b), which shows results corresponding to Fig. 8.8(a) when a back surface field is present.

8.5 TOP-LAYER LIMITATIONS

8.5.1 Dead Layers

In Section 8.3, it was shown that high effective recombination velocities at the top surface of the cell lead to an optimal design where the top diffused layer of the cell is made as thin as possible consistent with obtaining reasonable sheet resistivities. In cells developed in the 1960s for space use, junction depths beneath the surface were typically about 0.5 μm. As much phosphorus dopant as

---

[1] For high resistivities, minority carrier concentrations are likely to be comparable to majority carrier concentrations during cell operation. This not only complicates the calculation of series resistance but invalidates the analysis on which the expressions of this section are based. An expression for the dark current in this case of comparable simplicity to that implied by Eq. (8.31) is (Ref. 8.1):

$$I = \frac{qn_i W_P}{(\tau_e + \tau_h)} (e^{qV/2kT} - 1)$$

possible was diffused into this thickness to keep the sheet resistivity low. This created undesirable side effects.

From simple theoretical considerations (Ref. 8.2), a Gaussian or similar distribution of phosphorus is expected within the silicon after the high-temperature diffusion step from what is essentially an infinite source of phosphorus. Figure 8.9(a) is a sketch of the distribution of electrically active phosphorus, as typically measured after different diffusion times at fixed diffusion temperature. It clearly shows an upper limit to the amount of electrically active phosphorus. This limit equals the solid solubility of phosphorus in silicon at the diffusion temperature. Phosphorus in excess of this would be expected to be incorporated into phosphorus-rich precipitates. In regions of such excess phosphorus, minority-carrier lifetimes are dramatically reduced.

In a solar cell, regions of excess phosphorus would lie near the surface of the cell. This can produce a "dead layer" near the surface where light-generated carriers have very little chance of being collected because of the very low minority-carrier lifetimes. The collection probability corresponding to this situation is shown in Fig. 8.9(b). When this problem was clearly identified (Ref. 8.3), substantial modifications in the cell design were made to produce a high efficiency cell known as the "violet" cell. Much shallower junctions were used ($\sim 0.2 \, \mu$m) and the surface concentration of phosphorus was kept below the solid solubility limit, eliminating dead layers. This also reduced the sheet resistivity of the diffused layer, making it essential to use much closer spacing of the grid lines of the top contact.

8.5.2 High-Doping Effects

In the heavily doped top region of the cell, minority-carrier lifetimes would be expected to be low on several accounts. From Fig. 8.7, Auger recombination would result in low maximum values for lifetimes in this region. Additionally, the possibility of precipitates and defects in the crystal structure created by the high-temperature diffusion process would increase the number of centers for recombination via trapping levels. This would decrease lifetimes below the Auger limit.

Another effect of importance in heavily doped regions is the effective narrowing of the forbidden band gap of the semiconductor (Ref. 8.4). This will have a major influence on the effective value of the intrinsic concentration, n_i.

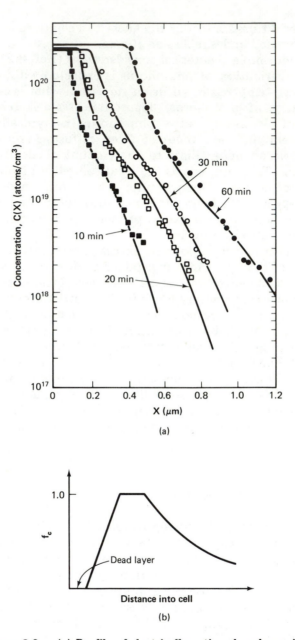

(a)

(b)

Figure 8.9. (a) Profile of electrically active phosphorus introduced into silicon for different diffusion times at fixed diffusion temperature. [After J. C. Tsai, *Proceedings of the IEEE 57* (1969), p. 1499, © 1969 IEEE.] (b) Collection probability as a function of distance into the cell for a diffused cell with a dead layer.

8.5.3 Contribution to Saturation Current Density

In Sections 8.3 and 8.4, contributions to the saturation current density, I_0, of the solar cell arising from the properties of the bulk region were discussed. The heavily doped top layer can also contribute significantly to this parameter.

Some general comments about the desirable properties of the top layer to minimize this contribution can be made. For example, recombination in this region and its associated surface has to be kept to a minimum. However, because of the multiplicity of effects that have to be taken into account in heavily doped regions, it is not straightforward to calculate how this is best accomplished from theoretical considerations. Experimentally, for diffused silicon cells, the minimum contribution to the saturation current density from this layer lies in the range of 1 to 3×10^{-12} A/cm^2. Ion-implanted top layers appear to be capable of giving slightly lower values (Ref. 8.5).

Regardless of how the substrate properties are chosen to minimize the substrate contribution to the saturation current density, the top-layer contribution will provide an upper limit to the maximum open-circuit voltage that a silicon cell of the type described can attain. This limit is in the range 600 to 630 mV at standard test conditions. The effect of this limit can be seen by referring to Fig. 8.8(a) and (b). It does not allow the full photovoltaic potential of the substrate material to be utilized. Alternative cell designs are described in Chapter 9 which overcome this limitation.

8.6 TOP-CONTACT DESIGN

One important area of cell design is the design of the top metal contact grid. This area becomes increasingly important as the size of individual cells increases. Figure 8.10 shows some of the diverse approaches to top contact design which have been used in terrestrial cells.

There are several power-loss mechanisms associated with the top contact. The loss due to lateral current flow in the top diffused layer of the cell has already been described. Additionally, there are losses due to the series resistance of the metal lines and the contact resistance between these lines and the semiconductor. Finally, there are the losses due to the shadowing of the cell by these lines.

In this section, the design of contacts for square or rectangular cells will be considered. Parallel techniques can be used for cells

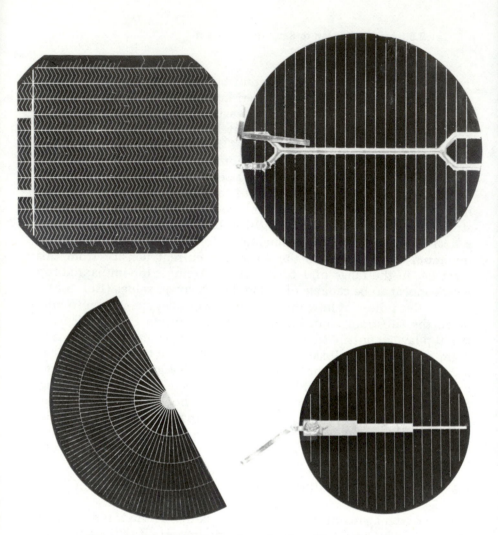

Figure 8.10. Collection of production silicon solar cells, showing diverse approaches to the design of the top contact to the cell.

of more general shape. For common contact designs, two types of metallic elements can be identified as indicated in Fig. 8.11(a). *Busbars* are relatively heavy areas of metallization directly connected to the external leads to the cell; *fingers* are finer elements which collect current for delivery to a busbar. As shown in Fig. 8.10, there can be more than one level of fingers in some cell designs. Fingers and busbars are usually either of constant width, linearly tapered, or have step changes in width.

154

A symmetrical contacting scheme such as in Fig. 8.11(a) can be broken down into unit cells as in Fig. 8.11(b). The maximum power output of this unit cell can be seen to be given by $ABJ_{mp}V_{mp}$, where AB is the area of the unit cell. J_{mp} and V_{mp} are the current density and voltage, respectively, at the maximum power point. The resistive losses in the fingers and busbars can be calculated using the integral approach as used to calculate the power loss in the top layer of the cell in Section 8.2.3. Normalized to the maximum output of the unit cell, the results are (Ref. 8.6)

$$p_{rf} = \frac{1}{m} B^2 \rho_{smf} \frac{J_{mp}}{V_{mp}} \frac{S}{W_F} \tag{8.32}$$

$$p_{rb} = \frac{1}{m} A^2 B \rho_{smb} \frac{J_{mp}}{V_{mp}} \frac{1}{W_B} \tag{8.33}$$

for the fractional resistive power loss in the fingers and busbars, respectively. ρ_{smf} and ρ_{smb} are the sheet resistivities of the contact metal layers for the fingers and busbars. In some cases these will be identical, whereas in others such as solder-dipped cells, where a thicker layer builds up over the wider busbar, ρ_{smb} will be smaller. The value of m is 4 if the respective element is linearly tapered and 3 if of uniform width. W_F and W_B are the average width of the fingers or busbar lying within the unit cell, and S is the pitch of the fingers as indicated in Fig. 8.11(b).

The fractional power losses due to shadowing by the fingers and busbar are

$$p_{sf} = \frac{W_F}{S} \tag{8.34}$$

$$p_{sb} = \frac{W_B}{B} \tag{8.35}$$

Neglecting current flows directly from the semiconductor to the busbar, the contact resistance loss is due solely to the fingers. The fractional power loss due to this effect is, as a general approximation

$$p_{cf} = \rho_c \frac{J_{mp}}{V_{mp}} \frac{S}{W_F} \tag{8.36}$$

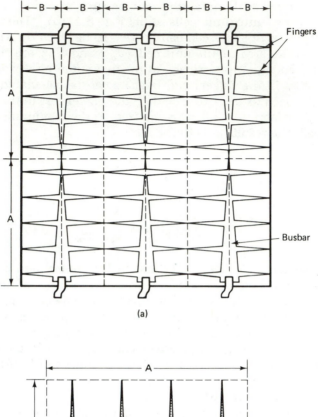

(a)

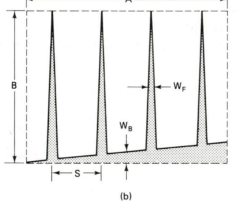

(b)

Figure 8.11. (a) Schematic of a top contact design showing busbars and fingers. Also shown is the symmetry in this particular design whereby the contact can be broken down into 12 identical unit cells. (b) Important dimensions of a typical unit cell.

156

where ρ_c is the specific contact resistance. Contact resistance losses are not generally an important consideration for silicon cells under one sun operation. The remaining loss is that due to the lateral current flow in the top layer of the cell. In normalized form, this is given by Eq. (8.23)

$$p_{tl} = \frac{\rho_s}{12} \frac{J_{mp}}{V_{mp}} S^2 \qquad (8.37)$$

where ρ_s is the sheet resistivity of this layer.

The optimum dimensions of the busbar can be found by summing Eqs. (8.33) and (8.35) and differentiation with respect to W_B to find its optimum value (Ref. 8.6). The result is that the optimum occurs when W_B is such that the resistive loss in the busbar equals its shadowing loss. This occurs when

$$W_B = AB \sqrt{\frac{\rho_{smb}}{m} \frac{J_{mp}}{V_{mp}}} \qquad (8.38)$$

with the minimum value of fractional power loss given by

$$(p_{rb} + p_{sb})_{min} = 2A \sqrt{\frac{\rho_{smb}}{m} \frac{J_{mp}}{V_{mp}}} \qquad (8.39)$$

This indicates that power losses are about 13% lower when a tapered busbar ($m = 4$) is used rather than a busbar of constant width ($m = 3$).

The design of the lowest level[2] of finger metallization is more complex, because this also determines the lateral loss in the top region of the cell and the contact resistance losses in the cell. Mathematically, the optimum occurs when the spacing between the fingers becomes very small to make the lateral loss negligible. The optimum is then given by the conditions

$$S \to 0 \qquad (8.40)$$

$$\frac{W_F}{S} = B \sqrt{\frac{\rho_{smf} + \rho_c m/B^2}{m} \frac{J_{mp}}{V_{mp}}} \qquad (8.41)$$

---

[2]If there is more than one level of finger metallization, the optimum dimensions of the higher levels can be found by regarding them as the busbars for the next lowest level. Note that in this case there will be a different unit cell for each level of metallization.

$$(p_{rf} + p_{cf} + p_{sf} + p_{tl})_{min} = 2B \sqrt{\frac{\sqrt{\rho_{smf} + \rho_c m /B^2}}{m} \frac{J_{mp}}{V_{mp}}} \quad (8.42)$$

In practice, it will not be possible to obtain this optimum performance. Each of the technologies for forming the top contacts previously discussed will place its own limit on how small W_F and hence S can be while still maintaining acceptable yields in a production environment.

In this case the optimum finger design can be realized by a simple iterative process. Regarding the finger width W_F as fixed at some minimum value by technological limitations, the optimum value of S corresponding to this can be found by successive approximations. At some trial value of S', the corresponding fractional power losses p_{rf}, p_{cf}, p_{sf}, and p_{tl} are calculated. A better approximation to the optimum value can then be calculated as[3]

$$S'' = \frac{S'(3p_{sf} - p_{rf} - p_{cf})}{2(p_{sf} + p_{tl})} \quad (8.43)$$

This process will quickly converge to a constant value corresponding to the optimum. A starting trial value for the optimum can be found by noting that the value of S calculated from Eq. (8.41) will be an overestimate. A value equal to half this should result in a stable iteration sequence.

The approach, as described above, will allow selection of the optimum dimensions of busbar and fingers once the gross features of a particular contact design have been specified. These gross features might be dictated by considerations other than optimized top-contact design such as the ease with which cells can be automatically interconnected. As a rule of thumb, the smaller the unit cell, the smaller the top-contact loss. Redundant contacts not only improve module reliability but also reduce top-contact losses by reducing the size of unit cells. If the sheet resistivity of the busbar is less than that of the fingers, it is best to use a design with long busbars and short fingers, as in the figure immediately preceding the Preface at the front of this book, provided that contact-resistance effects are small. In the case shown, the current-carrying section of the busbars

---

[3]This equation can be derived by differentiating the expression for the power loss $(p_{rf} + p_{cf} + p_{sf} + p_{tl})$ with respect to S. This derivative must equal zero for the optimum value of S. This optimum S is then found using the Newton iteration scheme (Ref. 8.7) for finding the roots of a nonlinear equation.

are the metal interconnect ribbons which extend the whole length of the cell. Even for rectangular cells, it may be worthwhile considering contact schemes other than the rectilinear ones discussed. For example, radial contact schemes such as illustrated in Fig. 8.12 can also give very low overall losses for rectangular cells.

It should be noted in passing that the equations described in this section are based on certain approximations (Ref. 8.8) regarding the size of the normalized power losses, the magnitude of resistive voltage drops, and the direction of current flow, particularly near intersections of fingers and busbars. Also, linear tapers may not be the optimum shape for busbars or fingers for differently shaped cells (Ref. 8.9). The effect of these second-order effects may be well worth checking for situations, such as in concentrating cells, where contact design is critical.

Problem: Design the top contact for a 10 × 10 cm silicon solar cell. This cell gives its maximum power output at a voltage of 450 mV and a current density of about 30 mA/cm². The sheet resistivity of the diffused layer in this cell is 40 Ω/□.

Two interconnections per cell are specified. The metallization is to be applied by plating and subsequent solder dipping. The finger width is specified as 150 μm. The sheet resistivity of the metallization is determined primarily by the solder layer, whose bulk resistivity is 15 μΩ-cm. The solder builds up to an average thickness of 42 μm over the fingers and 80 μm over the wider busbars. The specific contact resistance between the fingers and the semiconductor is 370 μΩ-cm².

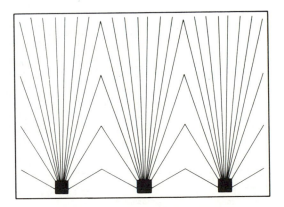

Figure 8.12. Radial contact metallization scheme for a rectangular cell.

Solution: In the terminology of this section,

$$J_{mp} = 0.03 \text{ A/cm}^2 \qquad V_{mp} = 0.45 \text{ V}$$

$$\rho_s = 40 \text{ }\Omega/\square \qquad \rho_c = 370 \text{ }\mu\Omega\text{-cm}^2$$

$$\text{Sheet resistivity of solder layers} = \frac{\text{bulk resistivity}}{\text{layer thickness}}$$

Therefore,

$$\rho_{smf} = 0.00357 \text{ }\Omega/\square \qquad \rho_{smb} = 0.00188 \text{ }\Omega/\square$$

Since $\rho_{smb} < \rho_{smf}$, a contact design with long busbars and short fingers is preferable. One suitable for two interconnections per cell is shown in Fig. 8.13. This breaks down into four unit cells, each with $A = 10$ cm, $B = 2.5$ cm.

The optimized dimensions of the busbar can be calculated from Eq. (8.38). Using a tapered busbar ($m = 4$), its optimum width per unit cell is

$$W_B = 10 \times 2.5 \left(\frac{0.00188 \times 0.03}{4 \times 0.45} \right) \text{ cm} = 0.14 \text{ cm}$$

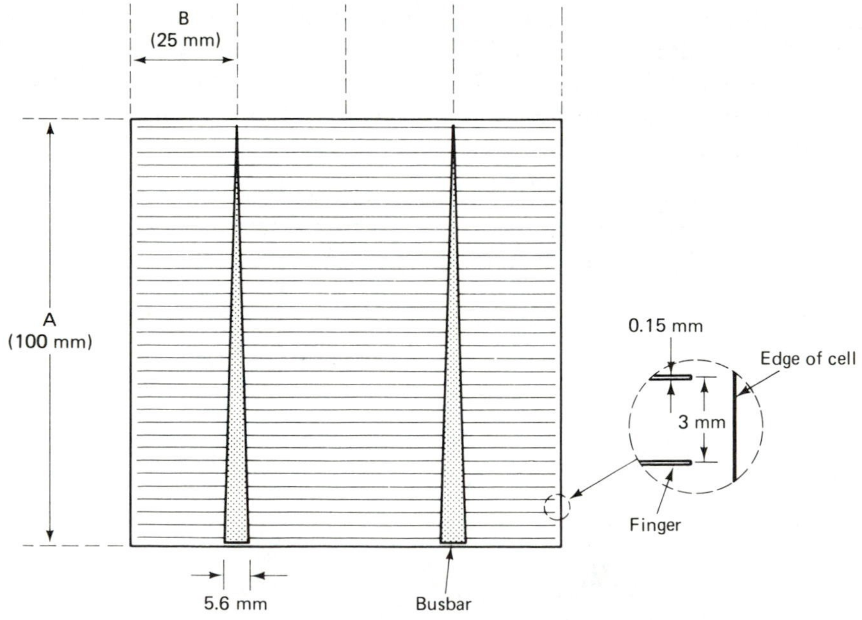

Figure 8.13. Contact scheme chosen for the design example. The shadowing loss is 10.6% and resistive losses in the contact and the diffused layer of the cell total 8.6%.

Since the actual busbar lies in two unit cells, the average busbar width is twice this. Hence, the busbar tapers from the thinnest value possible to a maximum width of 0.56 cm. The corresponding fractional power loss is, from Eq. (8.39),

$$p_{rb} + p_{sb} = 0.112$$

The finger width is specified as 150 μm ($W_F = 0.015$ cm). Assume that this is a result of technical limitations and the finger cannot be made any finer. With this constraint, a constant width ($m = 3$) finger will not be far below optimum. The optimum finger spacing S has to be found iteratively. A starting trial value is found by dividing the value given by Eq. (8.41) by 2. This gives

$$S = 0.3286 \text{ cm} \qquad p_{rf} = 0.0109 \qquad p_{cf} = 0.0005$$
$$p_{sf} = 0.0456 \qquad p_{tl} = 0.0240$$

Substituting these values into Eq. (8.43) gives the improved trial solution:

$$S = 0.2962 \text{ cm} \qquad p_{rf} = 0.0098 \qquad p_{cf} = 0.0005$$
$$p_{sf} = 0.0506 \qquad p_{tl} = 0.0195$$

Continuing this for another iteration gives

$$S = 0.2991 \text{ cm} \qquad p_{rf} = 0.0099 \qquad p_{cf} = 0.0005$$
$$p_{sf} = 0.0502 \qquad p_{tl} = 0.0199$$

Further iteration does not change the value of S, indicating that an optimum has been reached. The total fractional power loss due to the fingers and the top layer resistance is 0.080. The combined losses in the cell for this contact design is then 19.2% of the intrinsic output. The complete contact design is specified in Fig. 8.13.

8.7 OPTICAL DESIGN

8.7.1 Antireflection Coating

The principle of a quarter-wavelength antireflection coating is illustrated in Fig. 8.14. Light reflected from the second interface arrives back at the first interface 180° out of phase with that reflected from the first interface, canceling it out to some extent.

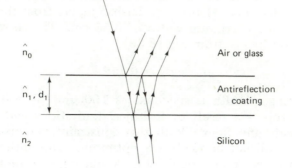

Figure 8.14. Interference effects created by a quarter-wavelength antireflection coating.

The expression for the fraction of the energy in a normally incident beam of light reflected from the surface of a material covered by a transparent layer of thickness d_1 is (Ref. 8.10)

$$R = \frac{r_1^2 + r_2^2 + 2r_1 r_2 \cos 2\theta}{1 + r_1^2 r_2^2 + 2r_1 r_2 \cos 2\theta} \qquad (8.44)$$

where r_1 and r_2 are given by

$$r_1 = \frac{n_0 - n_1}{n_0 + n_1} \qquad r_2 = \frac{n_1 - n_2}{n_1 + n_2} \qquad (8.45)$$

where the n_i represent the refractive indices of the different layers. θ is given by

$$\theta = \frac{2\pi n_1 d_1}{\lambda} \qquad (8.46)$$

When $n_1 d_1 = \lambda_0/4$, the reflectance has its minimum value:

$$R_{\min} = \left(\frac{n_1^2 - n_0 n_2}{n_1^2 + n_0 n_2} \right)^2 \qquad (8.47)$$

This is zero if the refractive index of the antireflection (AR) coating is the geometric mean of those of the materials on either side ($n_1^2 = n_0 n_2$). For a silicon cell in air ($n_{si} \approx 3.8$), the optimum refractive index is the square root of that of silicon (i.e., $n_{opt} \approx 1.9$). Figure 8.15

162

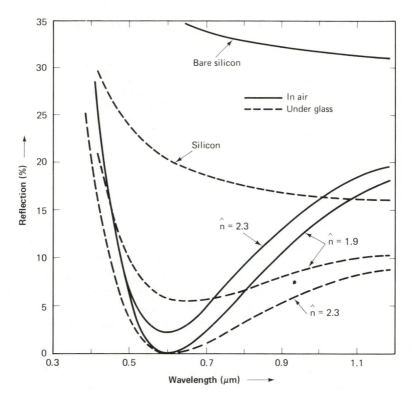

Figure 8.15. Percentage of normally incident light reflected from bare silicon and silicon with antireflection coatings of refractive indices of 1.9 and 2.3 as a function of wavelength. The thicknesses of the latter coatings are chosen to produce minimum reflection at 600 nm. The dashed lines show the effect of encapsulating the silicon under glass or material of a similar refractive index.

includes a curve showing the percentage of incident light reflected from silicon as a function of wavelength for an antireflection coating of this refractive index designed to produce minimum reflection at 600 nm. The weighted average of usable sunlight reflected from silicon can be kept to 10% as opposed to over 30% for bare silicon.

Cells are normally encapsulated under glass or embedded in a material of a similar refractive index to glass ($n_0 \sim 1.5$). This increases the optimum value of the index of the AR coating to about 2.3. Reflection from a cell with an AR coating of this refractive index is also shown in Fig. 8.15, both before and after encapsulation. The refractive index of a range of materials used in commercial solar cells is listed in Table 8.1. Apart from having the correct refractive index, the AR material must be transparent. It is usually deposited

163

Table 8.1. REFRACTIVE INDICES OF MATERIALS
USED IN SINGLE- OR MULTIPLE-LAYER
ANTIREFLECTION COATINGS

Material	Refractive index
MgF_2	1.3–1.4
SiO_2	1.4–1.5
Al_2O_3	1.8–1.9
SiO	1.8–1.9
Si_3N_4	~1.9
TiO_2	~2.3
Ta_2O_5	2.1–2.3
ZnS	2.3–2.4

as a noncrystalline or amorphous layer which prevents problems with light scattering at grain boundaries. Layers formed by vacuum evaporation generally tend to be absorbing at UV wavelengths. However, layers formed by such techniques as oxidizing or anodizing a deposited metal layer or layers deposited chemically tend to have a "vitreous" structure (amorphous with short-range order) which reduces this UV absorption (Ref. 8.11).

The use of multiple layers of AR coating materials can improve performance. The design of such coatings is more complex, but it is possible to reduce reflection over a broader band (Ref. 8.11). At least one manufacturer uses a two-layer coating in high efficiency cells to reduce the reflection of usable sunlight to 4%.

8.7.2 Textured Surfaces

Another approach previously mentioned to reduce reflection is to use textured surfaces. These are produced by etching the silicon surface with an etch that etches silicon much more rapidly in one direction through the crystal structure than another. This exposes certain planes within the crystal. The pyramids apparent in Fig. 7.6 are formed by the intersection of these crystal planes. In terms of Miller indices (Section 2.2), the silicon surface for textured cells is normally aligned parallel to a (100) plane and the pyramids are formed by the intersection of (111) planes. Dilute caustic soda (NaOH) solutions are a commonly used selective etch.

The angles of the pyramids are defined by the orientations of the crystal planes. These are such that incident light has at least two chances of being coupled into the cell. If 33% is reflected at each point of incidence, as is the case for normal incidence on bare silicon, the total reflected overall is 0.33×0.33, about 11%. If an antireflection coating is used, reflection of sunlight can be kept well below 3%. Even without an antireflection coating, the reflection when embedded in a material of refractive index similar to glass is only about 4%. Another desirable feature is that light is coupled into the silicon at an angle which ensures that it will be absorbed closer to the surface of the cell. This will increase the probability of its collection, particularly for the more weakly absorbed long wavelengths.

There are some disadvantages associated with the use of textured surfaces. One is that more care is required in handling (Ref. 8.12). A second is that such surfaces are much more effective in coupling light of all wavelengths into the cell, including unwanted infrared radiation of insufficient photon energy to create electron–hole pairs. This tends to make the cells "run hotter." Finally, top contact metallization has to run up and down the sides of the pyramids. If the height of the metal layer is less than or comparable to that of the pyramids (~ 10 μm), it follows that two to three times the metal must be used to maintain the same resistive loss as on a flat surface.

8.8 SPECTRAL RESPONSE

The spectral response of a cell was mentioned in Section 5.5. It is the output current under short-circuit per unit incident power in monochromatic light as a function of wavelength. Measurement of the spectral response can provide detailed information about the design parameters of any particular solar cell.

Monochromatic light causes electron–hole pairs to be generated in the semiconductor with a spatial distribution given by

$$G = (1 - R) \, \alpha N e^{-\alpha x} \qquad (8.48)$$

where N is the incident photon flux, R the fraction reflected, and α the absorption coefficient. For short wavelengths (UV light), α is large and light is absorbed quickly upon entering the semiconductor, as indicated in Fig. 8.16(a). Normal solar cells are not very effective at collecting light generated near the surface. If a quantum collection efficiency η_Q is defined as the number of electrons flowing in the external short-circuiting lead per incident photon in the monochromatic light, it will be very low for such UV light, as shown in Fig. 8.16(b). At intermediate wavelengths, α is smaller in value and a large proportion of the generated carriers are generated in regions where the collection probability is high. η_Q therefore increases.

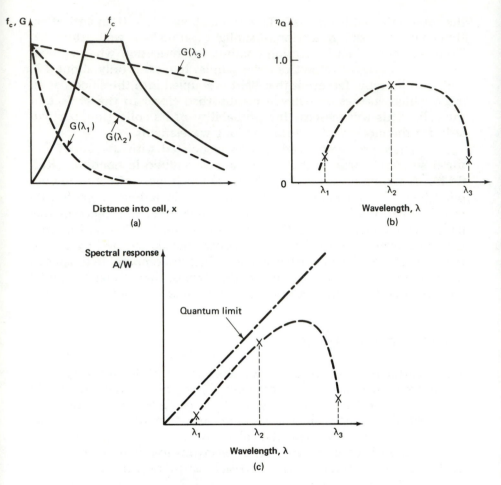

Figure 8.16. (a) Collection probability for a representative solar cell and, shown by the dashed lines, the carrier generation profiles for three different wavelengths of light. (b) Corresponding quantum collection efficiency, η_Q, as a function of wavelength. (c) Corresponding spectral sensitivity (A/W), also as a function of wavelength.

Long-wavelength light is absorbed very weakly and only a small proportion is absorbed in the active region of the cell. η_Q therefore decreases, dropping to zero once the photons have insufficient energy to create electron–hole pairs.

Another way of showing the spectral response apart from the quantum yield curve of Fig. 8.16(b) is to plot the sensitivity in A/W as a function of wavelength, as shown in Fig. 8.16(c). The quantum limit is also indicated. At short wavelengths, the cells cannot utilize all the energy in the photons, so the sensitivity is low, even if operating ideally.

For "conventional" cells (referring to the type of cell developed for use in space), the response at short wavelengths is poor due to the deep junction employed augmented by the absorption in the AR coating. At long wavelengths, the spectral response is determined by the diffusion length in the cell material. The violet cell was a shallow junction cell developed in the early 1970s. The design emphasis was on achieving good collection at UV wavelengths by using the shallower junction and an improved, less absorbing AR coating. Textured cells demonstrate improved sensitivity to light at all wavelengths due to reduced reflection.

A back surface field (BSF) increases the collection probability for carriers generated near the back contact and hence boosts the long-wavelength sensitivity. The back contact to the cell can be designed to be reflective to give the longer wavelengths a second chance at being absorbed. Such back surface reflectors (BSR) can give quite a boost to the performance of thin cells, as well as helping to maintain low operating temperatures.

8.9 SUMMARY

The design of silicon junction solar cells has evolved from the following considerations. The *p-n* junction has to lie close to the surface of the cell to give maximum current output. This can lead to problems due to the lateral resistance of this layer unless it is doped as high as practical. Excessive doping of this layer introduces effects that cause its electronic properties to be less than optimal.

The optimum substrate resistivity for a solar cell depends on whether a back surface field is present. If absent, the optimum resistivity is quite clearly defined at dopant levels in the range 10^{16} to 10^{17} cm^{-3}. If present, the optimum becomes less dependent on resistivity and is displaced to lower dopant levels.

In the design of the top contact to the cell, the critical parameters determining the associated power loss are the contact layout, the sheet resistivities of the contact metal layer and the diffused top layer of the cell, and the minimum width line permissible with the technology used to defined the contact geometry. A quarter-wavelength antireflection coating can increase the output current of a solar cell by 35 to 45%. Texturing the cell surface can produce even better performance, although there are some associated disadvantages.

EXERCISES

8.1. A particular silicon solar cell consists of a very thin, uniformly doped *n*-type layer on a *p*-type substrate. The surface recombination velocity along the surface of the *n*-type layer is very large. Assuming that minority-carrier diffusion lengths in the *n*-type layer are very *large* compared to the thickness of this layer, derive an expression for the probability of collection of minority carriers as a function of distance generated away from the surface in the *n*-type layer. (*Note*: Since the diffusion length is very large compared to the thickness of the *n*-type region, recombination in bulk of this region will be small and can be neglected compared to the rate of surface recombination.)

8.2. Calculate the sheet resistivity of the *n*-type layer in Exercise 8.1 given that the dopant concentration is 10^{18} cm^{-3} and the thickness of the layer is 0.5 μm.

8.3. Silicon solar cells of conventional structure are fabricated on *p*-type wafers 150 μm thick. Cells with a high recombination back contact give a short-circuit current of 2.1 A and an open-circuit voltage of 560 mV. Similar cells with a back surface field give a short-circuit current of 2.2 A. Given that the minority carrier diffusion length in the *p*-type region after processing in all cases is 500 μm, what would be the ideal value of the open-circuit voltage for the cells with a back surface field?

8.4. Design the top contact for a rectangular silicon solar cell 7.5 × 10 cm in size and calculate the total power loss of your design. The sheet resistivity of the diffused layer in this cell is nominally 60 Ω/□. The cell gives its maximum power at a voltage of 430 mV and a current density of 28 mA/cm^2 under bright sunshine. Three interconnections per cell are specified, all lying along the same side of the cell.

The metallization is to be applied by vacuum evaporation through a metal shadow mask. The minimum line width is specified as 180 μm. The metallization consists of a layer of titanium (0.12 μm) adjacent to the silicon separated by a thin layer of palladium (0.02 μm) from a 4-μm-thick layer of silver. The specific contact resistance to the silicon is 200 $\mu\Omega$-cm^2. The bulk resistivities of these metals are 48, 11, and 1.6 $\mu\Omega$-cm, respectively.

8.5. Consider a cell of given top contact geometry with metallization and cell parameters fixed. Show that the relative power loss attributable to this top contact increases as the cell dimensions increase.

REFERENCES

[8.1] J. G. FOSSUM et al., "Physics Underlying the Performance of Back-Surface-Field Solar Cells," *IEEE Transactions on Electron Devices ED-27* (1980), 785–791.

[8.2] A. S. GROVE, *Physics and Technology of Semiconductor Devices* (New York: Wiley, 1967), pp. 44–69.

[8.3] J. LINDMAYER AND J. F. ALLISON, *Conference Record, 9th IEEE Photovoltaic Specialists Conference, Silver Spring, Md.*, 1972, p. 83; also *Comsat Technical Review 3* (1972), 1.

[8.4] J. G. FOSSUM, F. A. LINDHOLM, AND M. A. SHIBIB, "The Importance of Surface Recombination and Energy-Bandgap Narrowing in *p-n* Junction Silicon Solar Cells," *IEEE Transactions on Electron Devices ED-26* (1979), 1294–1298.

[8.5] J. A. MINNUCCI et al., "Silicon Solar Cells with High Open-Circuit Voltage," *Conference Record, 14th IEEE Photovoltaic Specialists Conference, San Diego*, 1980, pp. 93–96; also *IEEE Transactions on Electron Devices, ED-27* (1980), 802–806.

[8.6] H. B. SERREZE, "Optimizing Solar Cell Performance by Simultaneous Consideration of Grid Pattern Design and Interconnect Configurations," *Conference Record, 13th IEEE Photovoltaic Specialists Conference, Washington, D.C.*, 1978, pp. 609–614.

[8.7] C. E. FROBERG, *Introduction to Numerical Analysis* (Reading, Mass.: Addison-Wesley, 1965), p. 19.

[8.8] A. FLAT AND A. G. MILNES, "Optimization of Multi-layer Front-Contact Grid Patterns for Solar Cells," *Solar Energy 23* (1979), 289–299.

[8.9] G. A. LANDIS, "Optimization of Tapered Busses for Solar Cell Contacts," *Solar Energy 22* (1979), 401–402; R. S. Scharlack, "The Optimal Design of Solar Cell Grid Lines," *Solar Energy 23* (1979), 199–201.

[8.10] E. S. HEAVENS, *Optical Properties of Thin Solid Films* (London: Butterworths, 1955).

[8.11] E. Y. WANG et al., "Optimum Design of Antireflection Coatings for Silicon Solar Cells," *Conference Record, 10th IEEE Photovoltaic Specialists Conference, Palo Alto*, 1973, p. 168.

[8.12] M. G. COLEMAN et al., "Processing Ramifications of Textured Surfaces," *Conference Record, 12th IEEE Photovoltaic Specialists Conference, Baton Rouge*, 1976, pp. 313–316.

Chapter

OTHER
DEVICE STRUCTURES

9.1 INTRODUCTION

The basis for photovoltaic action in a semiconductor device is an electronic asymmetry in the device structure. There are, of course, a large number of ways of creating this asymmetry apart from using silicon *p-n* junctions as considered in previous chapters. In this chapter, several alternative device concepts are outlined.

9.2 HOMOJUNCTIONS

Conventional silicon solar cells are *homojunctions*. The semiconductor on either side of the junction is the same, differing only in dopant type. In Chapter 8, the normal cell structure with a shallow homojunction parallel to the illuminated surface was described. Rather than cataloging all the alternative homojunction possibilities, three specific devices are described in this section that illustrate several different homojunction concepts.

The first is the high–low emitter (HLE) structure of Fig. 9.1(a), which overcomes some of the limitations of the conventional approach. This device differs in that the *p-n* junction is much deeper and the doping density more moderate on the top side of the junction. A "front surface field" is used at the top of the device. This approach overcomes the limitations imposed on the open-circuit voltage by the diffused top layer of the normal cell structure. Substantial improvements in open-circuit voltage have been reported using this approach (Ref. 9.1). The probability of collection of generated carriers throughout the device is indicated in Fig. 9.1(b). Carrier collection is not optimum, so current outputs tend to be less than with other designs.

A second device structure is the front-surface-field cell shown in Fig. 9.2(a). With this approach (Ref. 9.2) it is possible to bring both contacts out the rear of the cell. This eliminates the shadowing losses associated with the normal top contact and would make cells easier to interconnect. However, the processing required is more complex. The cell also has to be thin compared to the minority-carrier diffusion length to get full current output, which could pose handling difficulties for large-area cells.

A third approach is to use junctions perpendicular to the illuminated surface of the cell, the vertical junction cell. The most practical implementation of this approach has been the vertical multijunction (VMJ) cell of Fig. 9.2(b). Deep grooves are etched into the cell using an anisotropic etch (Ref. 9.3) and the cell subsequently diffused to give a combination of horizontal and vertical junctions. These vertical junctions ensure that carriers generated deep in the cell can be

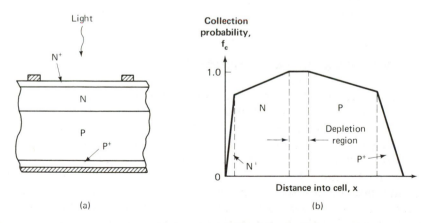

(a)   (b)

Figure 9.1. (a) Schematic diagram of the high–low emitter solar cell structure. (b) Corresponding collection probability as a function of distance from the surface of the cell.

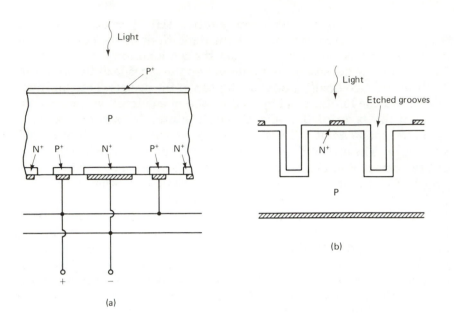

Figure 9.2. Other homojunction approaches to solar cell design:
(a) Front surface field cell.
(b) Vertical multijunction cell.

collected. To be most effective, the vertical junctions must be of the order of a diffusion length apart. In principle such a structure is more tolerant of small diffusion lengths, although the smaller the diffusion length the finer the required physical dimensions of the structure.

9.3 SEMICONDUCTOR HETEROJUNCTIONS

In semiconductor heterojunctions, the materials on either side of the junction are semiconductors but *different semiconductors.*

In Section 4.2, the energy-band diagram for the homojunction case was derived by a conceptual experiment where isolated *p*-type and *n*-type regions were brought into contact. Repeating this approach for the heterojunction case, the band diagrams of the isolated pieces of semiconductor are shown in Fig. 9.3(a). Three parameters are of interest. These are the *work function* (energy required to remove an electron located at the Fermi level from the semiconductor), the *electron affinity* (energy required to remove an electron located at the energy of the conduction-band edge), and the *semiconductor band gap.*

172

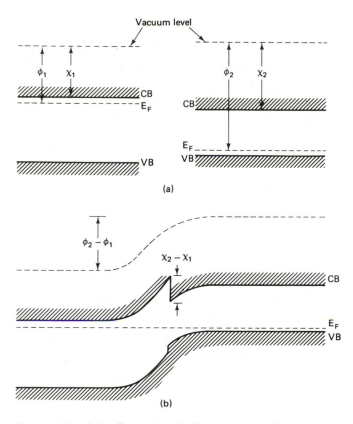

Figure 9.3. (a) Energy-band diagrams of isolated pieces of different semiconductors, one doped *p*-type, the other *n*-type. (b) Band diagram of the heterojunction formed by conceptually joining these pieces.

Not only are the doping levels different in the two isolated pieces, but the electron affinities, work functions, and band gaps are also different. When the two materials are joined together at thermal equilibrium, the Fermi level must be constant throughout the system, as indicated in Fig. 9.3(b). It follows that an electrostatic potential must build up across the device equal to the difference in work functions. The spatial distribution of this potential can be calculated in terms of charge stored in transition regions on either side of the junction in the same manner as in the homojunction case of Chapter 4. Additionally, there will be a discontinuity in the conduction-band edge at the junction equal to the difference in electron affinities and a corresponding discontinuity in the valence-band edge which also depends on the difference in band gaps. This is shown in Fig. 9.3(b). The *displacement vector* (ε ξ) will be continuous across ideal interfaces rather than the electric field as in the homojunction case.

Spikes in one of the bands as shown for the conduction band of Fig. 9.3(b) are undesirable for photovoltaic operation. Using the asymmetry concept of a solar cell as introduced in Section 4.1, the *n*-type region acts as a block for holes and the *p*-type as a block for electrons. A spike in the conduction band of the *n*-type region as in Fig. 9.3(b) also acts to block electron flow from the *p*-type to the *n*-type region. Hence, the spike will make it difficult for the *p*-type region to contribute to photocurrent. Such spikes can be avoided by a suitable combination of electron affinities and doping levels (Ref. 9.4).

For an ideal case with either small spikes or no spikes at all, the maximum efficiency of a heterojunction cell is bounded by the ideal efficiency of the smaller band-gap material. The reasons for considering heterojunctions are therefore practical reasons, rather than for fundamental efficiency advantages.

Up to the present point, nothing has been said about a very important practical consideration. For normal homojunctions, the same crystal structure continues right through the junction. For heterojunctions, this may not be possible because of gross differences in the crystal structure of the two semiconductors involved. From Fig. 9.4(a) it is apparent that for lattices of the same form but with a mismatch of lattice spacing, defects arise in the lattice structure. The

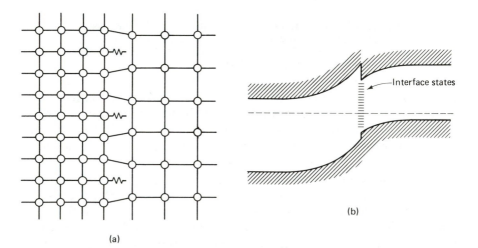

(a)

(b)

Figure 9.4.   (a) Defects due to the mismatch at the interface between two lattices of different lattice constants. (b) Corresponding defect states in the forbidden gap caused by this mismatch.

density of defects depends on the degree of mismatch between the lattices. Defects in the crystal lattice give rise to allowed energy levels within the forbidden gap, as illustrated in Fig. 9.4(b). These allowed energy levels lie within the depletion region, where they act as very efficient recombination centers. They can also provide sites for quantum mechanical tunneling processes for current transport from one side of the junction to the other. In any case, they act to degrade the performance of the solar cell. To produce heterojunctions with nearly ideal properties, it is essential to use semiconductors with nearly identical lattice structures.

9.4   METAL-SEMICONDUCTOR HETEROJUNCTIONS

When a metal and semiconductor are brought into contact, a potential drop occurs in the interfacial region to account for work function differences, as in the case of semiconductor heterojunctions. Because of differences in the availability of charge carriers in the metal and semiconductor, essentially all this potential drop occurs on the semiconductor side of the junction, as indicated in Fig. 9.5(a). This can give rise to a depletion region at the interface as in the *p-n* junction case. The metal acts similarly to very heavily doped semiconductor material from the point of view of its effects on the electrostatic properties of the depletion layer.

Metal–semiconductor contacts with such depletion regions are known as *Schottky diodes*. They have both rectifying and photovoltaic properties. The situation for *minority carriers* in the semiconductor

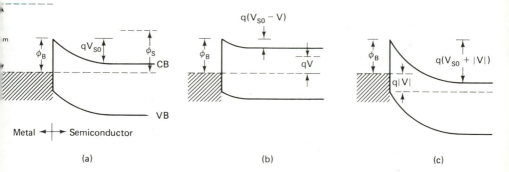

Figure 9.5.   Energy-band diagram at a metal–semiconductor heterojunction:
(a)   Zero bias.
(b)   Forward bias.
(c)   Reverse bias.

region is essentially the same as for the case of a *p-n* diode. For example, in the dark the excess minority-carrier concentration at the edge of the depletion region depends exponentially on the applied voltage, with an exponential decay into the bulk (Section 4.4 and 4.6). Minority-carrier flows give a similar contribution to the total diode current. For n-type semiconductor

$$J_{0h} = \frac{qD_h n_i^2}{L_h N_D} (e^{qV/kT} - 1) \tag{9.1}$$

The only impediment to majority-carrier flow between the metal and the semiconductor is the depletion-region potential barrier at the interface. The height of this barrier varies with applied voltage as shown in Fig. 9.5(b) and (c). This gives rise to a *thermionic emission* component of current given by (Ref. 9.5)

$$J_{0e} = A^* T^2 e^{-q\phi_B/kT} (e^{qV/kT} - 1) \tag{9.2}$$

where A^* is the effective Richardson constant (~ 30 to $120 \ A/cm^2/K^2$). The magnitude of this component depends primarily on the height of the barrier at the interface, ϕ_B. This majority-carrier component is normally much larger than the minority-carrier component of current, as indicated in Fig. 9.6(a). This extra component of current is *undesirable* from the point of view of photovoltaic energy conversion, because it acts to increase the dark saturation current of the diode and hence decrease the open-circuit voltage. This is indicated schemat-

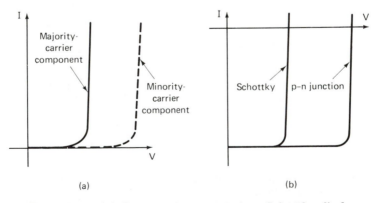

(a)　　　　　　　　　　　　　　　(b)

Figure 9.6.　(a) Current components in a Schottky diode in the dark. (b) Comparison of the illuminated characteristics of Schottky and *p-n* junction devices.

ically in Fig. 9.6(b). Therefore, *the larger the barrier* ϕ_B, the better the performance.

It may seem an easy matter to select the metal work function in such a way as to induce a very large barrier at the metal–semiconductor interface to overcome this problem. However, it is found experimentally that, for many semiconductors, the size of the barrier induced is independent of the work function of the metal. This is attributed to a very large number of interface states at the metal–semiconductor interface due to lattice mismatch and possibly the effect of contaminants on the semiconductor surface (Ref. 9.6). These act to clamp the potential in the surface region.

Even though Schottky diodes can be made very simply because no *p-n* junction formation is involved, their performance is limited by the presence of the additional "parasitic" current component over and above those associated with *p-n* junctions.

9.5 PRACTICAL LOW-RESISTANCE CONTACTS

From Section 9.4, since the size of the barrier at the metal–semiconductor interface can seldom be controlled by selecting an appropriate metal, the question arises: How is it ever possible to make a nonrectifying or low-resistance contact between a metal and a semiconductor? The answer follows if the width of the depletion region associated with the rectifying contact is examined. As the doping density in the semiconductor increases, this width will decrease.

A metal to heavily doped semiconductor contact is shown in Fig. 9.7. The depletion region can become so thin that carriers can pass through classically forbidden regions by quantum mechanical tunneling processes (Ref. 9.7). Essentially, this is due to the wavelike

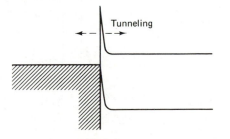

Figure 9.7. Contact between a metal and a heavily doped *n*-type semiconductor. The potential barrier in the latter regions is so thin that carriers can pass right through it by quantum mechanical tunneling processes.

nature of electrons which allows them to extend across such regions. Therefore, even though there is a barrier at the surface, carriers pass between the metal and semiconductor as if the thin barrier were not there. This gives a good low-resistance contact.

As a consequence, it is not difficult to make a good low-resistance contact to the heavily doped diffused layer of the cell. Electrical contact to the relatively lightly doped substrate region can be made by *alloying*, which leaves a heavily doped region near the interface. Alternatively, heavily damaged "as sawn" surfaces act sufficiently nonideally to prevent rectifying action.

9.6 MIS SOLAR CELLS

It has been seen that metal–semiconductor contacts act nonideally for many semiconductors in that barriers at the interface are not as strongly dependent on the work function of the metal as expected from simple theoretical considerations. By inserting a thin insulating layer between the metal and the semiconductor as indicated in Fig. 9.8(a), this nonideality can be removed. In such metal–insulator–semiconductor (MIS) devices, extreme metal work functions can produce extreme effects in the semiconductor.

For example, for the *p*-type device of Fig. 9.8(b), a low metal work function will cause a very large barrier to be induced at the semiconductor surface. If the insulating layer is very thin, carriers will be able to flow through it by *quantum mechanical tunneling*. The current

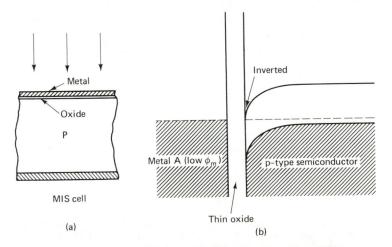

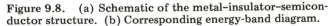

Figure 9.8. (a) Schematic of the metal–insulator–semiconductor structure. (b) Corresponding energy-band diagram.

that can flow through the insulator by this process increases exponentially with decreasing insulator thickness. The expression for the dark thermionic emission component of current given in Eq. (9.2) is then modified to

$$J_{0h} = P_h A^* T^2 e^{-q\phi_B/kT} (e^{q(V_{SO} - V_S)/kT} - 1) \qquad (9.3)$$

where V_S is the potential at the semiconductor surface and V_{S0} is its value at thermal equilibrium. P_h is the probability the particle has of tunneling through. The presence of the thin insulator will also tend to reduce the maximum rate at which minority carriers can pass between the metal and semiconductor. However, provided that the thin insulator layer is not too thick (typically less than about 20 Å), transport of these carriers will be limited by their smaller transport rate through the semiconductor. In this case, the situation for minority carriers will remain the same as for the case of a *p-n* junction diode, and the corresponding dark current flow will be given by (Section 4.6)

$$J_{0e} = \frac{qD_e n_i^2}{L_e N_A}(e^{qV/kT} - 1) \qquad (9.4)$$

Comparing the relative magnitudes of the two components of current flow for Schottky and MIS devices, the thermionic emission component of current can be suppressed in the latter by the fact that larger values of ϕ_B can be obtained, that the additional term P_h, the tunneling probability, can be very much smaller than unity, and that $(V_{S0} - V_S)$ in Eq. (9.3) may be smaller than the applied voltage V. From Fig. 9.6, such a suppression will increase the open-circuit voltage of the cell.

The ultimate situation is shown schematically in Fig. 9.9. In this case, the thermionic component of current has been suppressed to such low values that the *p-n* junction type of current flow given by Eq. (9.4) is left dominant. Hence, although structurally much different, MIS devices can be made electronically equivalent to ideal *p-n* junction diodes (Ref. 9.8).

In both Schottky and MIS solar cells, the top metal layer serves two functions, acting as both a contact and a barrier inducer. It is apparent from Fig. 9.8(a) that some way has to be found of getting the light through this layer. Two approaches are shown in Fig. 9.10. One is to use a metal layer thin enough (<100 Å) so that it is essentially transparent to light. This layer would have a high sheet resistivity, so a thicker contact grid would also be required as shown. A second approach is to use a grating structure essentially the same as the

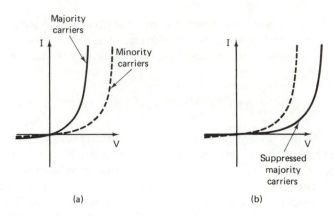

Figure 9.9. Comparison of the dark characteristics of (a) a Schottky and (b) an optimally designed MIS solar cell.

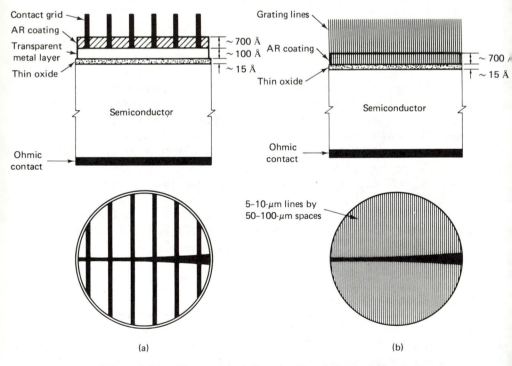

Figure 9.10. Two approaches to the design of the top contact of MIS solar cells:
(a) Transparent metal approach.
(b) Grating device.

top contact to conventional cells but very much finer. Carriers created by light absorbed between the grating lines have a good chance of reaching a nearby grating line before recombining. The situation can be further improved if it is possible to induce a layer of minority carriers along the surface electrostatically (Ref. 9.9). A third approach is to use a transparent conductor such as the oxides of tin, indium, zinc, or cadmium for the top contact. These conducting oxides are actually heavily doped semiconductors; hence, the resulting structure is known as the semiconductor–insulator–semiconductor (SIS) solar cell (Ref. 9.10). Such SIS devices are closely related to the semiconductor heterojunctions of Section 9.3.

The major advantage of the MIS approach is that the high-temperature junction diffusion step is completely eliminated. This allows the starting properties of the silicon material to be maintained as well as eliminating the nonidealities associated with the diffused layer. It was seen in Section 8.7 that these nonidealities place an upper limit on the open-circuit voltage of silicon cells. Using the grating MIS approach, very high open-circuit voltages have been reported for silicon solar cells (Ref. 9.8). Since current outputs are comparable and fill factors potentially higher, MIS cells have a potential efficiency advantage over diffused cells, owing primarily to this high output voltage.

9.7 PHOTOELECTROCHEMICAL CELLS

9.7.1 Semiconductor–Liquid Heterojunctions

If a semiconductor is brought into contact with an electrolyte, a potential barrier can be created at the surface of the semiconductor as for the other heterojunctions treated in this chapter. Minimal device processing is required to produce such semiconductor–liquid heterojunctions, which, nonetheless, have recorded energy-conversion efficiencies in excess of 12% (Ref. 9.11). The main problem with this approach has been the susceptibility of the semiconductor to photo-enhanced corrosion in this mode of operation (Ref. 9.12).

Combined with a conducting counter electrode, such "photoelectrical cells" can be used either to produce electricity or to produce hydrogen by the photoelectrolytic decomposition of water.

9.7.2 Electrochemical Photovoltaic Cells

In these devices, the liquid consists of a solution containing a species with an oxidized and a reduced state. If this species accepts an electron, it changes from an oxidized to a reduced state. Conversely, giving up an electron or accepting a hole oxidizes it. Such species are known as a *redox couples.*

In an electrochemical photovoltaic cell, the energy level within the electrolyte of this redox couple ideally lies near the energy level of the minority-carrier-band edge at the semiconductor. This is illustrated in Fig. 9.11(a) for the case of n-type semiconductor with a metallic counter electrode.

Under illumination, minority carrier holes generated in the semiconductor move to the interface with the electrolyte. They are transferred across this interface as indicated in Fig. 9.11(a) to the reduced form of the redox couple, which they oxidize:

$$\text{Red} + p^+ \to \text{Ox}^+ \tag{9.5}$$

At the counter electrode, electrons are transferred from the metal to the oxidized form of the redox couple, which they reduce:

$$\text{Ox}^+ + e^- \to \text{Red} \tag{9.6}$$

If a load is connected between the cell terminals, this will complete the electrical circuit and power will be supplied to it as in more conventional solar cells.

The electrolyte in these cells serves only to transfer charge between the metal and semiconductor. The device structure bears a

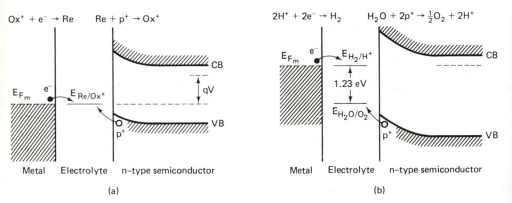

Figure 9.11. (a) Energy-band diagram of an electrochemical photovoltaic cell under illumination. The redox-couple level in the electrolyte allows charge transfer between the metal and the valence band in the semiconductor. (b) Energy-band diagram of a photoelectrolysis cell under illumination. Ideally, the device operates with the metal externally shorted to the rear of the semiconductor layer.

close relationship to the MIS structure of Section 9.6, as discussed further in Ref. 9.13.

9.7.3 Photoelectrolysis Cell

A cell very similar to the one described in Section 9.7.2 can also be used to produce chemical fuels by photoelectrolysis. The most common fuel produced has been hydrogen by the photoelectrolysis of water.

For an n-type semiconductor, reactions proceed as described in Section 9.7.2. An oxidation reaction occurs in the electrolyte near the semiconductor–electrolyte interface. A reduction reaction occurs near the counter electrode. The difference is that, unlike the case discussed in Section 9.7.2, the reacting species at each electrode are different. For the case of water decomposition, the reaction at the semiconductor electrode is (Ref. 9.14)

$$H_2O + 2p^+ \rightarrow \tfrac{1}{2}O_2 + 2H^+ \tag{9.7}$$

At the counter electrode,

$$2H^+ + 2e^- \rightarrow H_2 \tag{9.8}$$

The energy difference between the states associated with these reactions is 1.23 eV, as indicated in Fig. 9.11(b). This places a lower limit to the band gap of the semiconductor required for the reaction to proceed. All but semiconductors with much larger band gap than this have been found to corrode in this mode of operation (Ref. 9.12).

Titanium dioxide was the first semiconductor found to be stable in aqueous solution in this photoelectrolysis mode. However, it has a large band gap (3 eV) and so responds only to ultraviolet radiation. Solar-energy-conversion efficiencies are therefore low ($\sim$1%). In fact, this material absorbs so little sunlight that it is used as a "non-absorbing" antireflection coating on some commercial silicon cells! A small bias voltage (0.3 to 0.5 V) is required from an external power supply for this reaction to proceed for these TiO_2 devices. A material of smaller band gap with the required stability not requiring this bias is being sought.

9.8 SUMMARY

There is a wide range of photovoltaic device structural possibilities apart from the shallow homojunctions of previous chapters. Some alternative device structures have been described in the present chapter. Heterojunctions between two different band-gap semiconductors have no intrinsic efficiency advantage over homojunctions. As will be seen in Chapter 10, there are often technological advantages. Metal-semiconductor heterojunctions can be fabricated very simply but are intrinsically less efficient than homojunctions, owing to an additional parasitic component of current. However, the use of a metal–insulator–semiconductor (MIS) heterojunction structure can reduce or even completely eliminate this deficiency.

Heterojunctions formed between liquids and semiconductors also possess interesting photovoltaic properties. In a photovoltaic mode of operation, laboratory cells can be fabricated very simply and have demonstrated reasonable efficiencies. In a photoelectrolysis mode, sunlight is converted directly into chemical energy, usually in the form of stored hydrogen. This combined energy collection and storage mode gives rise to interesting possibilities provided that energy-conversion efficiencies can be drastically improved over those reported in the past.

EXERCISES

9.1.   (a) At room temperature, the dark saturation current density of homojunction cells made from material 1 is typically 10^{-8} A/m^2; the corresponding figure for cells from material 2 is 10^{-11} A/m^2. Which of these materials would you expect to have the smaller band gap?

      (b) A p-n heterojunction is formed between these two materials. Assuming the absence of current-limiting spikes at the metallurgical junction and similar lattice structures so that there is little mismatch at this junction, estimate the dark saturation current density expected from this heterojunction. Which material is most important in determining the open-circuit voltage of the cell?

      (c) Which material will set the upper limit to the number of electron–hole pairs generated in the heterojunction and hence the short-circuit current?

9.2.   (a) Find the ratio of the contributions made to the dark saturation current density of a Schottky diode solar cell by thermionic emission over the barrier at the metal–semiconductor interface and the diffusion of minority carriers through the semiconductor bulk at 300 K. The semiconductor is n-type silicon with the following parameters at 300 K: $N_D = 10^{22}$ m^{-3}, $D_h = 0.001$ m^2/s, $L_h = 10^{-4}$ m, and $n_i = 1.5 \times 10^{16}$ m^{-3}. The barrier height at the interface is 0.8 eV and the effective Richardson constant is 10^6 A/m^2/K^2.

      (b) If the cell gives a short-circuit current output of 300 A/m^2 under bright sunshine (1 kW/m^2), calculate the ideal value of the open-circuit voltage and cell efficiency.

9.3. A MIS cell is made with a structure similar to the Schottky cell of Exercise 9.2. Assume that the device achieves optimal performance by suppression of the magnitude of the thermionic emission component of current well below that due to the diffusion of minority carriers. Calculate the value of the open-circuit voltage and efficiency and compare to the results of Exercise 9.2.

9.4. Under bright sunshine (1 kW/m^2), a photoelectrolysis cell based on titanium dioxide generates hydrogen at the diode surface. The cell requires a bias voltage of 0.4 V to operate and draws a current of 7 A/m^2 of cell area from the bias supply. The potential power which can be extracted from the hydrogen is 1.48I, where I is the cell current and 1.48 is the voltage equivalent of the heat of combustion of hydrogen. What is the efficiency of conversion of sunlight for this cell?

REFERENCES

[9.1] F. A. LINDHOLM et al., "Design Considerations for Silicon HLE Solar Cells," *Conference Record, 13th IEEE Photovoltaic Specialists Conference, Washington, D.C.*, 1978, pp. 1300-1305; also C. T. SAH et al., *IEEE Transactions on Electron Devices ED-25* (1978), 66.

[9.2] O. VAN ROOS AND B. ANSPAUGH, "The Front Surface Field Solar Cell, a New Concept," *Conference Record, 13th IEEE Photovoltaic Specialists Conference, Washington, D.C.*, 1978, pp. 1119-1120.

[9.3] J. WOHLGEMUTH AND A. SCHEININE, "New Developments in Vertical Junction Silicon Solar Cells," *Conference Record, 14th IEEE Photovoltaic Specialists Conference, San Diego*, 1980, pp. 151-155.

[9.4] W. D. JOHNSTON, JR., AND W. M. CALLAHAN, *Applied Physics Letters 28* (1976), 150.

[9.5] S. M. SZE, *Physics of Semiconductor Devices* (New York: Wiley, 1969), p. 378.

[9.6] IBID, p. 372.

[9.7] B. SCHWARTZ, ed., *Ohmic Contacts to Semiconductors* (New York: Electrochemical Society, 1969).

[9.8] M. A. GREEN, F. D. KING, AND J. SHEWCHUN, "Minority Carrier MIS Tunnel Diodes and Their Application to Electron- and Photo-voltaic Energy Conversion: Theory and Experiment," *Solid State Electronics 17* (1974), 551-572; R. B. GODFREY AND M. A. GREEN, "655 mV Open Circuit Voltage, 17.6% Efficient Silicon MIS Solar Cells," *Applied Physics Letters 34* (1979), 790-793.

[9.9] P. VAN HALEN, R. E. THOMAS, AND R. VAN OVERSTRAETEN, "Inversion Layer Silicon Solar Cells with MIS Contact Grids," *Conference Rec-*

ord, *12th IEEE Photovoltaic Specialists Conference, Baton Rouge,* 1976, pp. 907-912.

[9.10] R. SINGH, M. A. GREEN, AND K. RAJKANAN, "Review of Conductor-Insulator-Semiconductor (CIS) Solar Cells," *Solar Cells 3* (1981), 95-148.

[9.11] A. HELLER, B. A. PARKINSON, AND B. MILLER, "12% Efficient Semiconductor-Liquid Junction Solar Cell," *Conference Record, 13th IEEE Photovoltaic Specialists Conference, Washington, D.C.,* 1978, pp. 1253-1254.

[9.12] H. P. MARUSKA AND A. K. GHOSH, "Photovoltaic Decomposition of Water at Semiconductor Electrodes," *Solar Energy 20* (1978), 443-458.

[9.13] S. KAR et al., "On the Design and Operation of Electrochemical Solar Cells," *Solar Energy 23* (1979), 129-139.

[9.14] A. J. NOZIK, "Electrode Materials for Photoelectrochemical Devices," *Journal of Crystal Growth 39* (1977), 200-209.

Chapter

10

OTHER SEMICONDUCTOR MATERIALS

10.1 INTRODUCTION

The semiconductor material that has been the focus of attention in earlier chapters has been single-crystal silicon. There is a wide range of other semiconductor material capable of producing solar cells of acceptable efficiencies (Ref. 10.1). No attempt is made in this chapter to catalog such material. Rather, the structure and properties of solar cells made on some of the more developed alternatives to single-crystal silicon are discussed. This will serve to indicate considerations of importance for more general materials.

10.2 POLYCRYSTALLINE SILICON

Techniques for preparing polycrystalline silicon are, in general, less critical than those required to produce single-crystal silicon. The purity of the starting silicon used to produce the polycrystalline material must still be similar to that used in single-crystal material to get acceptable photovoltaic performance. But what other properties must polycrystalline silicon possess to produce reasonable solar cells?

The important regions in polycrystalline cells are the boundaries between the grains. An electrostatic barrier tends to develop on either side of the grain boundary (Ref. 10.2) similar to that developed in the metal–semiconductor heterojunction of Section 9.3. This tends to block majority carrier flows, acting essentially as a large series resistance. It makes the columnar grain structure of Fig. 10.1(b), where grains extend from front to back of the cell, more desirable than the case of Fig. 10.1(a), where they do not. Being defects in the crystal structure, grain boundaries introduce allowed levels into the forbidden gap of the semiconductor material and act as very effective recombination centers. As such, they may be considered as "sinks" for minority carriers. Just as minority carriers generated within a diffusion length of the junction of a solar cell are collected by the junction, so those generated within about the same distance of a grain boundary are attracted to the boundary and recombine. Such carriers do not contribute to the output current of the cell. As a consequence, it follows that the lateral dimensions of grains within polycrystalline material must be large compared to minority carrier diffusion lengths to avoid significant loss in current output (Ref. 10.3). Another deleterious effect that can be attributed to grain boundaries is their ability to provide shunting paths for current flow across the p–n junction. These conducting paths may develop from the preferential diffusion of dopants down grain boundaries during the junction formation step, as indicated in Fig. 10.1(c). A large density of precipitates at the grain boundaries would also be expected to contribute to these shunting paths.

Silicon, being a weakly absorbing indirect-band-gap semiconductor (Section 3.3.2), requires large diffusion lengths of the order of 0.1 mm for good photovoltaic performance. To avoid a signifi-

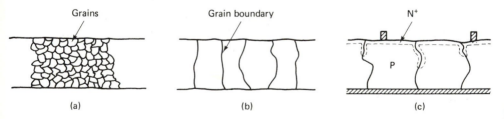

Grains   Grain boundary   N⁺

(a)   (b)   (c)

Figure 10.1. (a) Fine-grained polycrystalline material. (b) Polycrystalline material with columnar grains extending across the wafer thickness. (c) Preferential diffusion of dopant impurities down grain boundaries during cell processing.

cant loss of photocurrent due to recombination at grain boundaries, grains must have lateral dimensions much larger than this, of the order of a few millimeters. Since cells are normally only a fraction of a millimeter thick, such large grain size makes the columnar condition of Fig. 10.1(b) relatively easily satisfied. Moreover, the total length of grain boundaries per unit area of cell decreases as the grain size increases, decreasing the importance of shunting effects arising from these boundaries.

Such large grain sizes are much larger than those normally associated with polycrystalline material and the term *semicrystalline* is more appropriate. Figure 10.2 shows a wafer of such semicrystal-

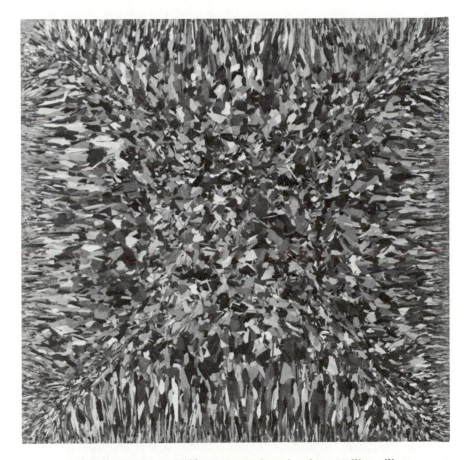

Figure 10.2. A 10 × 10 cm wafer of polycrystalline silicon sliced from an ingot formed by a casting process. Solar cells of 10% efficiency can be routinely produced on such wafers.

line silicon sliced from a cubic ingot of this material. In 1976, it was shown that solar cells of efficiencies in excess of 10% could be fabricated on such material (Ref. 10.4). Larger-grained material has since been reported to give efficiencies over 14% (Ref. 10.5). Solar modules based on such semicrystalline silicon are now available commercially.

10.3 AMORPHOUS SILICON

Conditions for preparing amorphous silicon are even less critical, in principle, than those for preparing polycrystalline silicon. An amorphous material differs from crystalline material in that there is no long-range order in the structural arrangement of the atoms. For example, silicon atoms in amorphous silicon will generally be connected to four other silicon atoms. The angles between the bonds joining these atoms as well as the bond lengths generally will be similar to those in crystalline material, but small deviations quickly result in a complete loss of long-range order.

Elemental amorphous silicon does not appear to possess any interesting photovoltaic properties in itself. Without the constraints of periodicity, it is difficult for each silicon atom to be linked up with four others. This gives rise to microvoids within the structure of the material with associated unsatisfied or "dangling" bonds. Combined with the nonperiodic arrangement of the atoms, this creates large densities of allowed states right across the normally forbidden band gap. These make it impossible to effectively "dope" the semiconductor or obtain reasonable carrier lifetimes.

However, it was reported in 1975 (Ref. 10.6) that amorphous silicon films produced by the glow discharge decomposition of silane (SiH_4) could be doped to form $p-n$ junctions. These films contained hydrogen (created when the SiH_4 decomposes) as a reasonable proportion of the total atoms within the material (5 to 10%). It has been postulated that the role of hydrogen is to saturate dangling bonds on the internal microvoids of the film and at other defects in the structure as indicated in Fig. 10.3. This would reduce the density of states within the forbidden gap and allow the material to be doped.

Using amorphous silicon–hydrogen alloys (a-Si:H alloys) prepared in this way, a 5.5% efficient solar cell was reported in 1976 (Ref. 10.7). This was a very small area device based on a MIS structure, but it did bring attention to the possibilities of the approach. Much larger area $p-n$ junction and MIS devices have since been fabricated with efficiencies approaching this value (Ref. 10.8). The a-Si:H

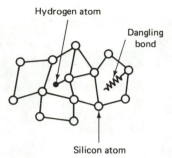

Hydrogen atom

Dangling bond

Silicon atom

Figure 10.3. Schematic diagram of the structure of amorphous silicon, showing how dangling bonds arise and how these may be passivated by hydrogen.

alloys display significantly larger band gaps than crystalline silicon and also are much more strongly absorbing. This means that films of the order of 1 μm thick are all that are required. These films can be deposited onto a variety of substrates. Doping levels can be controlled by the presence of small quantities of gases containing the desired dopant during the deposition. Results indicate that minority carrier diffusion lengths in such material are very small, much less than 1 μm. As a consequence, the depletion region forms most of the active carrier collecting volume of the cell. Series resistance of the bulk regions of the cells tends to be a problem. However, reduction of this resistance when the cell is illuminated (*photoconductive effect*) compensates to some extent.

Since the cells are so easily deposited, it is not difficult to form several interconnected cells on the one substrate. This has the advantage that the size of any individual cell can be kept small, eliminating the requirement for a metal contact grid (Ref. 10.9). The first commercial products based on amorphous silicon appeared in 1980. These utilized small cells interconnected on the one substrate as shown in Fig. 10.4(a). These produce the required voltage and current to power the consumer products indicated in Fig. 10.4(b). These cells were over 3% efficient in sunlight but *matched the performance of single-crystal silicon cells in indoor fluorescent lighting*. The search is under way in research laboratories around the world to find a material based on amorphous silicon which will repeat this feat outdoors.

A development in this area is the use of amorphous silicon layers with both hydrogen and fluorine incorporated into the structure (Ref. 10.10). These a-Si:F:H alloys have been produced by the glow discharge decomposition of SiF$_4$ in the presence of hydrogen. This approach has been reported to lead to more desirable

191

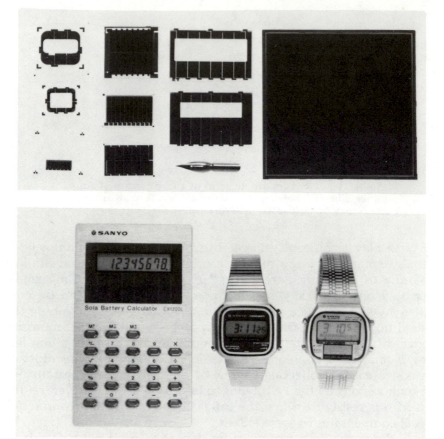

Figure 10.4. (a) First commercial amorphous silicon solar cell arrays designed for use in consumer products. (b) Calculator and watches powered by these arrays. (Photographs courtesy of Sanyo Electric Co. Ltd.)

properties for photovoltaic action, in particular a reduced density of states within the forbidden gap (Ref. 10.10).

10.4 GALLIUM ARSENIDE SOLAR CELLS

10.4.1 Properties of GaAs

Gallium arsenide (GaAs) is an example of a *compound semiconductor*. It has a similar crystalline structure to silicon (Fig. 2.3) except that every second atom is different (either Ga or As). GaAs is also a direct-band-gap semiconductor (Section 3.3.1). This means that sunlight is absorbed very quickly after entering it. It also means that minority carrier lifetimes and diffusion lengths are much smaller than in silicon. These differences result in different cell design concepts.

GaAs has a well-developed technology because of the commercial interest in this material for light-emitting diodes and semiconductor injection lasers. One aspect of this technology has been the utilization of alloys of GaAs and AlAs. The latter material is an indirect-band-gap semiconductor ($E_g \approx 2.2$ eV) with a lattice structure similar to GaAs, with the lattice spacing nearly identical (only 0.14% mismatch). Alloys of GaAs and AlAs, normally written $Ga_{1-x}Al_xAs$, have a lattice spacing and band gap intermediate between GaAs and AlAs. Because of the good match of lattice spacing, heterojunctions between GaAs, AlAs, and their alloys can have low densities of interfacial states and, as a consequence, nearly ideal properties. This adds an additional degree of flexibility to the design of photovoltaic devices.

Because of its nearly ideal band gap (Fig. 5.2) and its developed technology, the most efficient solar cells ever reported have been based on GaAs. Terrestrial efficiencies of over 22% have been reported under AMl sunlight, substantially higher than the corresponding figure of 18% for silicon. However, there are some disadvantages with the use of GaAs as a solar cell material. The limited resources of gallium (Ref. 10.11) will ensure that GaAs is always an expensive material. This is offset by the fact that GaAs cells are ideal for use in systems that concentrate sunlight (Chapter 11). The amount of material required for a given power output is reduced as a consequence. The direct band gap of GaAs also means that light is absorbed very quickly after entering this material. Therefore, layers only a few microns thick are all that are required, further reducing material requirements. A second disadvantage is the toxic nature of arsenic. The environmental consequences of deploying large solar systems based on toxic materials would have to be carefully examined (Ref. 10.12).

10.4.2   GaAs Homojunctions

Because light is absorbed very quickly upon entering a direct-band-gap semiconductor such as GaAs, problems with high surface-recombination velocities are even more serious for the conventional homojunction structure than for the case of silicon. Prior to 1978, efficiencies reported for GaAs homojunctions had been moderate.

The technique used for reducing surface-recombination effects for silicon was to make the top layer of the homojunction thin compared to the average depth where photons are absorbed (Section 6.2.2). The same technique also works for GaAs, although in this case, the layer has to be an order of magnitude thinner. Efficiencies in excess of 20% have been reported for an N^+PP^+ solar cell structure of Fig. 10.5(a) in which the top N^+ layer was only 450 Å thick (Ref. 10.13).

For fabricating cells based on GaAs, there is an emphasis on different processing techniques than those described for conventional silicon cells. Rather than forming doped layers by diffusing impurities into the GaAs, the more common technique is to chemically build up layers incorporating the required dopant density. These layers added to the device continue the crystal structure of the underlying substrate and are known as *epitaxial layers.* They are formed by heating the substrate in the presence of chemicals in either the vapor or the liquid phase which contain the materials it is desired to deposit.

The structure of Fig. 10.5(a) can be formed by starting with a heavily doped P^+ substrate, epitaxially growing a more lightly doped P layer of a few microns and then growing a thin heavily doped N^+ layer. Anodic oxidation of some of this layer to form the antireflection coating helps keep the recombination velocity at the surface of this layer to a minimum (Ref. 10.13).

10.4.3 $Ga_{1-x}Al_xAs/GaAs$ Heteroface Cells

An alternative technique used to overcome the limitations of high surface-recombination velocities for direct-band-gap GaAs is to use the *heteroface* structure of Fig. 10.5(b). Because of the nearly identical structure of GaAs and alloys with AlAs, it is possible to build up an epitaxial layer of $Ga_{1-x}Al_xAs$ on the surface of a homojunction cell. If the parameter, x, is about 0.8, this layer will have a large indirect band gap. There will be little absorption of sunlight in passing through such a layer. In essence, it will act as a "window layer," allowing light through to the underlying cell as indicated in Fig. 10.5(d). Since the good lattice match to the substrate causes few interface states at the heteroface junction, the window layer also passivates the surface of the underlying GaAs.

This device structure has resulted in the best single cell efficiencies yet reported, with values in excess of 22% in terrestrial sun-

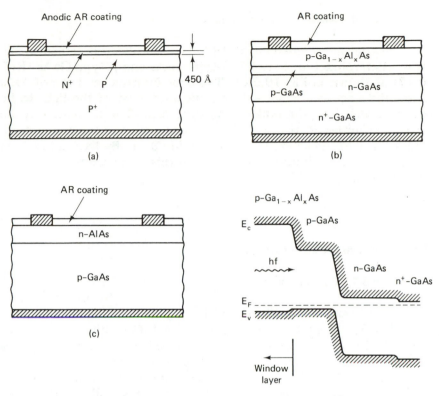

Figure 10.5. Schematic diagrams of different approaches to the design of GaAs solar cells:

(a) GaAs homojunction.
(b) $Ga_{1-x}Al_xAs/GaAs$ heteroface cell.
(c) AlAs/GaAs heterojunction.
(d) Energy-band diagram corresponding to the heteroface cell shown in part (b).

light (Ref. 10.14). The processing sequence is to start with an n-type substrate of GaAs and epitaxially grow the overlying p-type $Ga_{1-x}Al_xAs$ layer from the liquid phase. The same processing step also dopes the top portion of the substrate p-type by the diffusion of p-type impurities. Obtaining reliable low-resistance contacts to the $Ga_{1-x}Al_xAs$ layer can present difficulties that will be particularly important for cells to be used in concentrated sunlight. This problem can be avoided either by etching through this layer in the contact area and making contact to the underlying p-type GaAs (Ref. 10.15) or by building up a heavily doped p-type GaAs layer on top of it in the contact area (Ref. 10.16).

195

10.4.4 AlAs/GaAs Heterojunctions

Efficiencies in excess of 18% have also been obtained with a "true" heterojunction between n-type AlAs and p-type GaAs (Ref. 10.17) as shown in Fig. 10.5(c). The large indirect band gap of AlAs causes it to act as a window layer, allowing most of the light to be absorbed well into the bulk of the solar cell. The mismatch in the electron affinities of AlAs and GaAs causes a spike in the conduction-band energy of the heterojunction (Section 9.3). By having the AlAs heavily doped, the undesirable effects of this spike can be minimized (Ref. 10.17).

10.5 Cu₂S/CdS SOLAR CELLS

10.5.1 Cell Structure

CdS cells have a history of development dating back to 1954 (Ref. 10.18), about the same year as the first results appeared for diffused silicon cells. Since then there have been several attempts to produce a commercial solar cell based on this material.

The striking feature of these cells is the ease with which they can be fabricated. Because fine-grained polycrystalline CdS is adequate as a substrate material, this opens up a large number of ways of preparing such substrates. Vacuum evaporation and spraying methods seem the most promising.

CdS cells are normally made by a process known as the *Clevite process*. CdS is vacuum-evaporated onto a metal sheet or a metal-covered plastic or glass sheet. The CdS has to be deposited to a thickness of only 20 μm or so. The diameter of the crystallites in this layer may be approximately 5 μm. The layer is then dipped in a cuprous chloride solution (80 to 100°C) for 10 to 30 s. This substitutes Cu for Cd in a thin surface region about 1000 to 3000 Å thick, giving a Cu_2S/CdS heterojunction. A grid contact is then deposited. Figure 10.6(a) shows the completed cell structure. The Cu_2S layer can extend several microns down grain boundaries, as shown. A diagram showing the energy-band relationships is given in Fig. 10.6(b). The Cu_2S material is p-type with a band gap of 1.2 eV, whereas the CdS is n-type with a gap of 2.3 eV.

Efficiencies of cells produced in this way have been in excess of 9% (Ref. 10.19), with about 5% obtainable in pilot production.

196

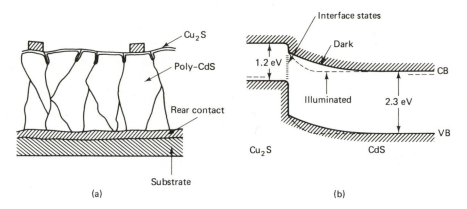

Figure 10.6. (a) Schematic diagram of the Cu_2S/CdS thin-film solar cell. (b) Corresponding energy-band diagram in the dark and when illuminated.

10.5.2 Operating Characteristics

Cu_2S/CdS cells perform almost anomalously well considering their inherent simplicity. However, the mechanisms responsible for their good performance are not nearly as well characterized as for Si or GaAs cells. Their operation can be described only by introducing concepts additional to those described in connection with the latter cell types.

The response of Cu_2S/CdS cells involves several nonlinearities. The most obvious demonstration of this is the fact that their current–voltage characteristics under illumination can cross over their dark characteristics, as demonstrated in Fig. 10.7(a). In addition, the open-circuit voltage of these cells and their fill factors depend on the spectral content of the light source, not just the light-generated current density, as also shown in Fig. 10.7(a). The cell capacitance also increases by a large factor (in the range 10 to 100) when the cells are illuminated. The spectral response of the cells shown in Fig. 10.7(b) also depends strongly on the intensity (and also the spectral composition) of the bias light. Although all the foregoing effects are generally observed, their magnitudes vary widely with different fabrication conditions.

The capacitance results indicate that the width of the depletion region contracts under illumination as illustrated in Fig. 10.6(b). A possible explanation of this (Ref. 10.20) is that trapping levels are created in the area of the depletion region by copper impurities which diffuse into this region during the formation of the Cu_2S layer

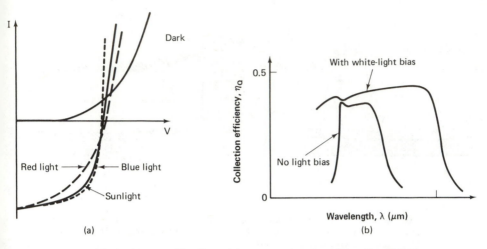

(a)

(b)

Figure 10.7. Nonlinearities associated with Cu₂S/CdS solar cell operation:
(a) Crossover of dark and illuminated curves and dependence upon the spectral content of the illumination.
(b) Enhancement of the spectral response by the application of bias light. (After Ref. 10.20, © 1978 IEEE.)

or subsequent heat treatment. Under illumination light-generated holes may be trapped in these levels. This would act to augment the positive charge due to donor impurities in this n-type region and hence decrease the depletion-layer width, as described mathematically in Eq. (4.4). This would also increase the value of the electric field strength in this region, as also indicated by Eq. (4.4).

This also allows the spectral response of the cells to be explained. There is likely to be a large number of allowed states in the forbidden gap at the Cu₂S/CdS interface due to lattice mismatch. These are indicated in Fig. 10.6(b). They act as efficient recombination centers. However, it is possible to show that the efficiency of such recombination centers decreases in the presence of large fields. Carriers are swept past these centers in such cases so rapidly that the probability of recombination is decreased. Most of the light-generated current comes from the thin Cu₂S top layer with its 1.2-eV band gap. Relatively little comes from the wider-band-gap (2.3 eV) CdS layer (Ref. 10.21). With no bias light, fields at the interface are relatively small, so that collected carriers have a good chance of recombining and the spectral response of the cells is poor, as shown in Fig. 10.7(b). With bias light, fields become large, recombination decreases, and the spectral response improves.

The sensitivity of the characteristics to different wavelengths of light can then be related to the relative ability of different wavelengths to cause traps near the interface to be occupied by holes. However, the explanation given above is only one of many possible explanations of the characteristics, and even if it were largely correct, additional mechanisms would be required to be incorporated into it to describe all the observed experimental features.

10.5.3   Advantages and Disadvantages of Cu_2S/CdS Cells

The prime advantage of Cu_2S/CdS cells is the ease with which they can be fabricated on a variety of supporting substrates, making the cells extremely well suited for large-scale automated production. The cells themselves can be produced very inexpensively.

The major disadvantages of these cells are their low efficiency and their lack of the inherent stability possessed by silicon cells. With low efficiencies, the costs of other parts of a system become more important because the area of solar cells required for a given output increases. Balance-of-system costs, such as those due to site preparation, support structures, and wiring, can dominate system costs to such an extent that even if the cells were free, it would be cheaper to use higher-efficiency cells of higher cost. As a rule of thumb, 10% module efficiency is the lowest that can probably be tolerated for cost-effective large-scale generation of photovoltaic power.

The same consideration applies to the encapsulation costs of the cells. This is compounded by the fact that Cu_2S/CdS cells are inherently less stable than other cell types that have been developed and hence require more careful encapsulation if they are to have comparable operating life.

A number of degradation modes have been demonstrated by CdS cells (Ref. 10.22): (1) under high humidity, (2) at high temperatures ($> 60°C$) in air, (3) when illuminated at high temperatures, and (4) when the load voltage exceeds 0.33 V.

Moisture absorbed in the cell creates additional traps which act to decrease the short-circuit current. This is a reversible process, and the original current can be retained with appropriate heat treatment. If the devices are heated above $60°C$ in air, irreversible changes in this current can occur, attributed to conversion of the Cu_2S to mixtures of CuO and Cu_2O by reaction with oxygen and moisture. Even if air is not present, illumination at such temperatures can decrease the efficiency. This is attributed to a light-activated phase change in the Cu_2S layer, where some is converted to Cu_xS with

$x < 2$. This change in stoichiometry lowers the efficiency considerably. Operation of the cells at voltages in excess of 0.33 V can cause a light-activated change of Cu_2S to CuS and Cu. The Cu can form fine filaments, effectively shunting the junction.

It is maintained that these degradation modes can be eliminated by minor changes in the cell fabrication method and by hermetic encapsulation of the cells. From the point of view of the widespread application of photovoltaics, the CdS technology suffers from the less-than-infinite reserves and the toxicity of Cd.

10.6 SUMMARY

There is a large range of materials that may prove viable alternatives to the single-crystal silicon which forms the basis of the present solar cell industry. Some of the more developed of these have been described in this chapter.

Gallium arsenide is a semiconductor with a well-developed technology and a nearly ideal band gap for solar photovoltaic energy conversion. The most efficient solar cells made to date have been based on this material. Homojunction, heteroface, and heterojunction cells have all been shown capable of high efficiency, overcoming the more severe constraints of surface-recombination velocity in direct-band-gap material. The high cost of this material is its major limitation.

Copper sulfide–cadmium sulfide cells can be fabricated very simply on small-grained polycrystalline cadmium sulfide. Although potentially a very low cost technology, the difficulty of achieving 10% efficiency on a production basis combined with stringent encapsulation requirements to prevent degradation may prevent its widespread use.

Polycrystalline silicon has the disadvantage that large grain sizes are required. This eliminates many low-cost preparation methods. Even with large-grained semicrystalline material, there are some advantages over single-crystal silicon which have resulted in commercial cells based on such technology. The most promising silicon thin-film technology is that based on amorphous silicon alloys. Very rapid laboratory and commercial progress has been made with this technology since its desirable features were identified.

EXERCISES

10.1. A thin layer of polycrystalline silicon is deposited on a supporting metallic substrate. The grains in this layer have the columnar structure of Fig.

10.1(b), with the lateral dimensions of the grains equal to the layer thickness. A *p–n* junction is formed close to the surface of this layer with negligible preferential diffusion of dopants down grain boundaries. Assume that the grain boundaries act as infinitely large sinks for minority carriers and that the recombination velocity at the metal–semiconductor interface at the rear of the cell is also very large. Approximating each grain as a cube, calculate the maximum probability that a minority carrier generated right at the center of the volume defined by a grain has of contributing to the short-circuit current of the cell. (*Note:* Under short-circuit, the *p–n* junction is a very attractive region for minority carriers acting as another sink. The maximum collection probability will occur when the minority carrier diffusion length is much larger than the grain dimensions. This corresponds to the case where no recombination occurs within the internal volume of the grain.)

10.2. For the heteroface solar cell of Fig. 10.5(b), sketch the form of the rate of generation of electron–hole pairs as a function of distance beneath the cell surface.

10.3. A certain technology produces 10% efficient solar modules at a cost of $1 per peak watt output under bright sunshine (1 kW/m^2). In a particular application, those balance-of-system costs that depend on the area of the array deployed amount to $80 per m^2. Assuming that other costs are identical in each case, at what price would 5% efficient modules produced by a second technology have to sell to give similar overall system costs?

REFERENCES

[10.1] A. L. FAHRENBRUCK, II–VI Compounds in Solar Energy Conversion," *Journal of Crystal Growth 39* (1977), 73–91; A. M. Barnett and A. Rothwarf, "Thin-Film Solar Cells: A Unified Analysis of Their Potential," *IEEE Transactions on Electron Devices ED-27* (1980), 615–630; S. Wagner and P. M. Bridenbaugh, "Multicomponent Tetrahedral Compounds for Solar Cells," *Journal of Crystal Growth 39* (1977), 151–159; M. Schoijet, "Possibilities of New Materials for Solar Photovoltaic Cells," *Solar Energy Materials 1* (1979), 43–57.

[10.2] J. G. FOSSUM AND F. A. LINDHOLM, "Theory of Grain-Boundary Intragrain Recombination Currents in Polysilicon *p–n* Junction Solar Cells," *IEEE Transactions on Electron Devices ED-27* (1980), 692–700.

[10.3] H. C. CARD AND E. S. YANG, "Electronic Processes at Grain Boundaries in Polycrystalline Semiconductors under Optical Illumination," *IEEE Transactions on Electron Devices ED-24* (1977), 397–402.

[10.4] H. FISCHER AND W. PSCHUNDER, "Low Cost Solar Cells Based on Large Area Unconventional Silicon," *Conference Record, 12th IEEE Photovoltaic Specialists Conference, Baton Rouge, 1976*, pp. 86–92.

[10.5] J. LINDMAYER AND Z. C. PUTNEY, "Semicrystalline versus Single Crystal Silicon," *Conference Record, 14th Photovoltaic Specialists Conference, San Diego*, 1980, pp. 208–213.

[10.6] W. E. SPEAR AND P. G. LeCOMBER, *Solid State Communications 17* (1975), 1193.

[10.7] D. E. CARLSON et al., "Solar Cells Using Schottky Barriers on Amorphous Silicon," *Conference Record, 12th IEEE Photovoltaic Specialists Conference, Baton Rouge*, 1976, pp. 893–895.

[10.8] D. E. CARLSON, "An Overview of Amorphous Silicon Solar-Cell Development, *Conference Record, 14th IEEE Photovoltaic Specialists Conference, San Diego*, 1980, pp. 291–297.

[10.9] J. J. HANAK, "Monolithic Solar Cell Panel of Amorphous Silicon," *Solar Energy 23* (1979), 145–147; Y. Kuwano et al., "A Horizontal Cascade Type Amorphous Si Photovoltaic Module," *Conference Record, 14th IEEE Photovoltaic Specialists Conference, San Diego*, 1980, pp. 1408–1409.

[10.10] A. MADAN, S. R. OVSHINSKY, AND W. CZUBATYJ, "Some Electrical and Optical Properties of a-Si:F:H Alloys," *Journal of Electronic Materials 9* (1980), 385–409.

[10.11] H. J. HOVEL, *Solar Cells*, Vol. 11, Semiconductor and Semimetal Series (New York: Academic Press, 1975), pp. 217–222.

[10.12] T. L. NEFF, "Comparative Social Costs and Photovoltaic Prospects," *Conference Record, 13th IEEE Photovoltaic Specialists Conference, Washington, D.C.*, 1978, pp. 1001–1003.

[10.13] J. C. C. FAN AND C. O. BOZLER, "High-Efficiency GaAs Shallow-Homojunction Solar Cells," *Conference Record, 12th IEEE Photovoltaic Specialists Conference, Washington, D.C.*, 1978, pp. 953–955.

[10.14] J. M. WOODALL AND H. J. HOVEL, *Applied Physics Letters 30* (1977), 492.

[10.15] R. SAHAI et al., "High Efficiency AlGaAs/GaAs Concentrator Solar Cell Development," *Conference Record, 13th IEEE Photovoltaic Specialists Conference, Washington, D.C.*, 1978, pp. 946–952.

[10.16] H. A. VANDER PLAS et al., "Performance of AlGaAs/GaAs Terrestrial Concentrator Solar Cells," *Conference Record, 13th IEEE Photovoltaic Specialists Conference, Washington, D.C.*, 1978, pp. 934–940.

[10.17] W. D. JOHNSTON, JR. AND W. M. CALLAHAN, "Vapor-Phase-Epitaxial Growth, Processing and Performance of AlAs-GaAs Heterojunction Solar Cells," *Conference Record, 12th IEEE Photovoltaic Specialists Conference, Baton Rouge*, 1976, pp. 934–938.

[10.18] F. A. SHIRLAND, "The History, Design, Fabrication and Performance of CdS Thin Film Solar Cells," *Advanced Energy Conversion 6* (1966), 201–222.

[10.19] J. A. BRAGAGNOLO et al., "The Design and Fabrication of Thin-Film CdS/Cu$_2$S Cells of 9.15 Percent Conversion Efficiency," *IEEE Transactions on Electron Devices ED-27* (1980), 645–651.

[10.20] A. ROTHWARF, J. PHILLIPS, AND N. CONVERS WYETH, "Junction Field and Recombination Phenomena in CdS/Cu$_2$S Solar Cell," *Conference Record, 13th IEEE Photovoltaic Specialists Conference, Washington, D.C.*, 1978, pp. 399–405.

[10.21] J. A. BRAGAGNOLO, "Photon Loss Analysis of Thin Film CdS/Cu$_2$S Photovoltaic Devices," *Conference Record, 13th IEEE Photovoltaic Specialists Conference, Washington, D.C.*, 1978, pp. 412–416.

[10.22] Reference 10.11, pp. 195–198.

Chapter
11

CONCENTRATING SYSTEMS

11.1 INTRODUCTION

One possible approach to reducing the cost of photovoltaic power even with present cell technology is to reduce the area of cell required for a given power output by concentrating the sunlight. In this way the cost of the system can be displaced from the cells to the cost of the concentrating elements and sun tracking system if required.

In general, the higher the concentration ratio, the smaller is the range of angles of light rays that the system will accept. Once the concentration ratio becomes greater than about 10, the system can utilize direct sunlight only and must track the sun in its path across the sky. The higher the concentration ratio, the more accurately the sun must be tracked. The finite range of angles reaching the earth from the sun due to its finite size determines the maximum possible concentration ratio (about 45,000).

Concentrating the sunlight on a cell will also tend to increase its operating temperature, which will decrease its efficiency. Passive

204

cooling (fins, etc.) is adequate for concentration ratios up to about 50. Active cooling is required for higher concentration ratios. *Total energy systems* which use both the photovoltaic energy and the thermal energy collected by the cooling system are entirely feasible.

11.2 IDEAL CONCENTRATORS

The geometrical concentration ratio, C, is defined as the ratio of the aperture area of the system to the active cell area. As mentioned previously, this ratio is closely related to the range of angles, θ_m, which the concentrating system can accept. From thermodynamic arguments it is possible to derive relationships between the maximum possible concentration ratio and the acceptance angle. For systems that concentrate light arriving at different angles within the range of acceptance angles by an equal amount, the maximum concentration ratio is given by (Ref. 11.1)

$$C_{m(2D)} = \frac{1}{\sin{(\theta_m/2)}} \qquad (11.1)$$

for a *two-dimensional* or *linear concentrator* as indicated in Fig. 11.1(a), and

$$C_{m(3D)} = \frac{1}{\sin^2(\theta_m/2)} \qquad (11.2)$$

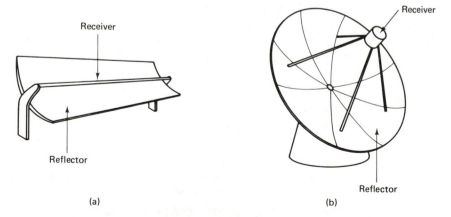

(a)   (b)

Figure 11.1. (a) Two-dimensional or linear concentrator. (b) Three-dimensional or point concentrator.

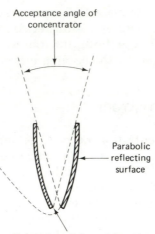

Acceptance angle of
concentrator

Parabolic
reflecting
surface

Focus of parabola on right

Figure 11.2. Schematic diagram of a nonimaging compound parabolic concentrator.

for a *three-dimensional* or *point-concentrating* system, as shown in Fig. 11.1(b). The range of angles of rays in direct sunlight due to the finite size of the sun's disk is about $\frac{1}{2}°$ (9.4 mrad). This gives a maximum possible concentration of sunlight of 45,000 for a point-concentrating system.

The ideal performance of conventional concentrators such as focusing parabolas and lenses falls short of the limits of Eqs. (11.1) and (11.2) by a factor of 2 to 4 (Ref. 11.2). The first concentrator identified as having an ideal performance equal to these limits was the nonimaging compound parabolic concentrator (CPC) shown in Fig. 11.2. It consists of two parabolic reflectors aligned with their focal points lying in the positions shown.

11.3 STATIONARY AND PERIODICALLY ADJUSTED CONCENTRATORS

For stationary concentrators and those whose orientation is adjusted either daily or seasonally, it is clearly desirable to have as large an acceptance angle as possible consistent with obtaining high concentration. As an example, consider the case of a trough concentrator with its longitudinal axis lying in the east–west direction. The important variation in the direction of the sun's rays comes about due

206

to changes in the elevation of the sun. It can be shown that the solar elevation lies within $\pm 36°$ of the plane of its path at the equinoxes for at least 7 h on any day of the year. Hence, an optimally designed stationary trough concentrator would concentrate sunlight for at least 7 h each day if its acceptance angle was $72°$. The maximum concentration obtainable is $1/\sin(72°/2)$, a very modest 1.7. Higher concentration ratios are possible if shorter minimum collection times are used or if the concentrator is designed to be periodically tilted.

Argonne National Laboratories fabricated small photovoltaic modules based on CPCs in 1976. One set of modules used parabolic reflectors; a second used total internal reflection within a solid acrylic block with the CPC shape (Ref. 11.3). These required seasonal adjustment to give concentration ratios of 7 to 9.

The preceding discussion indicates that the concentration obtainable for stationary concentrators is relatively low (considerably less than 3), whereas periodic adjustment increases achievable concentration ratios to about 12. The concentration obtained with a truly stationary concentrator may seem so marginal as to not warrant the extra complexity. However, one desirable feature, particularly for asymmetrical concentrators (Ref. 11.4), is that they can be used to boost the winter output of solar systems relative to the summer output. For stand-alone systems (Chapter 13), such a concentrator could not only decrease the cell area required but could also decrease the amount of energy storage required and the severity of cyclic drains on this storage.

One advantage of relatively low-concentration-ratio systems (<5 times) is that it may be possible to utilize cells fabricated in large volume for nonconcentrating applications and achieve the associated economies of scale. Higher-concentration-ratio systems require design changes in cells.

A novel approach to nontracking concentrators is the luminescent concentrator (Ref. 11.5). A sheet of glass or plastic is doped with a luminescent substance. A photocell is attached along one edge of the sheet and the other three edges are made reflective, as depicted in Fig. 11.3. Incident sunlight is absorbed by the dopant and then reemitted by luminescence in a narrow-wavelength range. A large fraction of the emitted light is trapped within the sheet either by total internal reflection or by reflection from the reflective edges until it reaches the attached photocell. The concentration obtainable with this system is not bounded by the limits mentioned previously. All angles of incidence can be accepted and maximum concentration is limited by practical considerations such as absorption of the emitted light in the sheet.

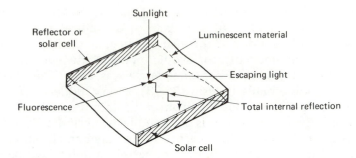

Figure 11.3. Luminescent concentrator. Absorbed sunlight is reemitted by luminescence with a large proportion trapped within the sheet by total internal reflection. The trapped light eventually reaches a solar cell.

11.4 TRACKING CONCENTRATORS

The mainstream of activity in the area of concentrating photovoltaic systems involves systems that concentrate sunlight by a factor greater than 20 and track the sun. Several diverse approaches to the design of such systems have been explored.

This diversity is brought out by two approaches to the design of a 10-kW subunit installed at Sandia Laboratories during 1978 and 1979. The first, shown in Fig. 11.4(a), uses a parabolic trough to concentrate sunlight onto a secondary concentrator and then onto the cell (Ref. 11.6). The overall geometrical concentration ratio is 25. This arrangement relaxes the precision required for the primary concentrator, although it does mean that two reflections are required, reducing the light reaching the cells to a maximum of 78% of that incident. Note the rather substantial heat sinking to keep the cells passively cooled. A 10-kW array mounted on a circular track for azimuth tracking is shown in Fig. 11.4(b).

An alternative approach shown in Fig. 11.5 uses refractive effects. A Fresnel lens has several advantages in this type of application. Note that the lens not only concentrates the sunlight but also provides an enclosure for the cell. The system shown uses a quad lens of this type to focus light onto cells mounted on a heat sink. With this design, the heat-sink area can be as large as the system aperture, keeping the cells reasonably cool even at the design concentration ratio of 40. A 2.2-kW array using this approach is shown (Ref. 11.7). A 350-kW$_p$ system based on units similar to this was installed in Saudi Arabia during 1980 and 1981. At the time, this was the largest photovoltaic system deployed.

208

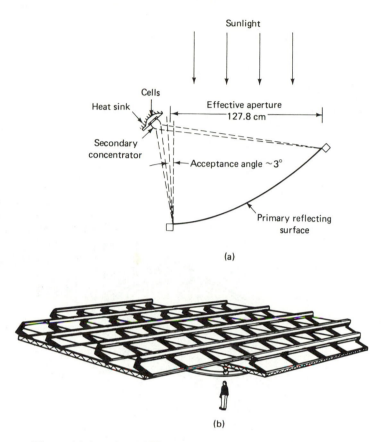

Figure 11.4. A 10-kW concentrating photovoltaic system based on a parabolic trough concentrator:

(a) Concentrating elements and cell mounting.
(b) Overall system mounted on a circular track.

(After Ref. 11.6, © 1978 IEEE.)

11.5 CONCENTRATOR CELL DESIGN

The ideal efficiency of cells *held at fixed temperature* increases with increasing concentration ratio. This is because the short-circuit current increases linearly with intensity, the open-circuit voltage increases logarithmically, and the fill factor increases with open-circuit voltage. The major difficulty in realizing this efficiency advantage is the increased importance of series resistance losses at high current densities. Since the cell efficiency determines the area of concentrating elements required for a given power output, it is critically important to achieve as high an efficiency as possible.

209

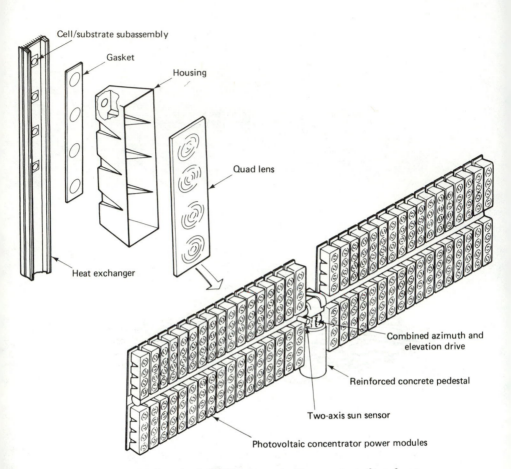

Figure 11.5. A 2.2-kW concentrating system based on Fresnel lenses. The lenses not only concentrate the sunlight but form part of the cell housing. In this design, the area of the heat sink is comparable to the system aperture. (After Ref. 11.7, © 1978 IEEE.)

The following suggestions could be made to minimize the resistance of a solar cell: (1) use of a low-resistivity substrate with a back surface field region for low bulk and contact resistive losses, (2) minimization of the sheet resistivity of the normally diffused thin top layer, (3) use of a top contact design with a fine finger pattern to minimize resistive losses due to lateral current flows, and (4) use of thick metal contact layers to reduce resistive losses in fingers and busbars.

All these steps are normally taken to produce present concentrator cells. Lower-resistivity substrates have been used than are

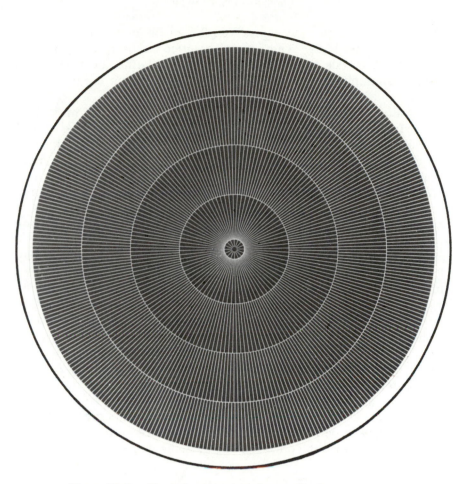

Figure 11.6. Typical concentrating cell for a point-focusing system. (Cell courtesy of Applied Solar Energy Corporation.)

used in normal cells.[1] The sheet resistivity of the diffused layer is kept reasonably low. However, striving for too low a value can give rise to the performance degradation effects described in Chapter 7. Each technology for defining the top contact places a lower limit to how fine it can be made. This limit also depends on the thickness of the metal required in the top layer. As a rule of thumb for the

[1] Because of 'high injection' effects, the series resistance of cells on high resistivity substrates with a back surface field can be kept low under some conditions. This provides an alternative approach to making concentrator cells. See the footnote in Section 8.4 and the reference therein.

dimensions involved, such lines can be made only about half as thick as they are wide. The normal practice has been to deposit the top contact metals by vacuum evaporation, define the pattern required using photolithographic methods, and then build up the lines as thick as possible by electroplating silver.

For concentrating, cells are normally designed to have their top surface only partially illuminated. For example, for the typical contact design of Fig. 11.6 for point-concentrating systems, the area not covered by the busbar around the edge is the design area. Efficiencies are based on the amount of light reaching this design area *rather than on the total cell area.* The form of the efficiency variation as a function of concentration ratio at fixed cell temperature is shown in Fig. 11.7. The general trend is an increase in cell efficiency with increasing concentration ratio for low values of this ratio and a decrease in cell efficiency at high values. The peak efficiency may occur at concentration ratios anywhere between 20 and several hundred suns. The increase at low concentrations is due to the logarithmic increase of voltage output with increasing current density. At high current density, series resistance losses become important and decrease the efficiency by decreasing the fill factor.

Peak efficiencies in excess of 20% have been reported for silicon concentrating cells (Ref. 11.8) compared to about 25% for cells based on GaAs (Ref. 11.9). Actual operating efficiencies tend to be somewhat lower than this because cells in concentrating systems are likely to operate at relatively high temperatures due to increased power densities. Optical losses reduce system efficiencies further. Designers of concentrating systems would be very pleased

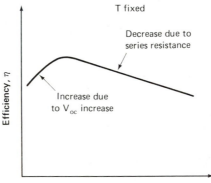

Figure 11.7. Schematic diagram showing the typical variation in efficiency of a solar cell held at fixed temperature as the concentration ratio is increased.

if they were able to design systems where 85% of the light striking the system aperture reached the cells.

11.6 ULTRA-HIGH-EFFICIENCY SYSTEMS

11.6.1 General

The energy-conversion efficiency of the photovoltaic section of a concentrating system is a key parameter in determining the cost of the system. It determines the system aperture for a given output. In the following sections, concepts that result in exceptionally high energy-conversion efficiencies will be explored. This ability to obtain very high efficiencies is one aspect of concentrating systems that makes them a distinctly different alternative to flat-plate modules.

11.6.2 Multigap-Cell Concepts

Selecting the optimum band gap for the solar cell material can be regarded as a trade-off between choosing the band gap narrow enough so that not too many photons are lost because they have insufficient energy to create electron–hole pairs, yet wide enough so that not too many waste their energy by creating electron–hole pairs with energy well in excess of the band gap.

A more efficient system can be produced if the low-energy photons in sunlight are directed to solar cells made from narrow-band-gap semiconductors in which they can be utilized. The high-energy photons are directed to wide-band-gap cells, where their energy is not dissipated by creating electron–hole pairs with energies grossly in excess of the band gap.

Two concepts for directing light to cells of appropriate band gap are shown in Fig. 11.8. The first, known as *spectrum splitting*, uses spectrally sensitive mirrors to direct the light to appropriate cells. The second, the *tandem cell* approach, uses a series of cells stacked on top of each other with the widest-band-gap material uppermost. Low-energy photons will pass through the stack until they reach a cell of low enough band gap to utilize them. Since both schemes involve extra complexity compared to a single cell, the multicell concept is best suited to systems using high concentration ratios.

The maximum efficiency obtainable with this approach depends on the number of different band gap cells used. Table 11.1

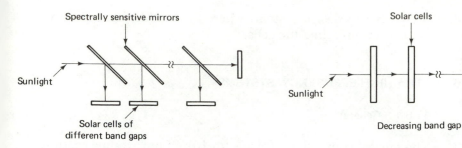

Figure 11.8. Multigap-cell concepts:
(a) Spectrum-splitting approach.
(b) Tandem-cell approach.

shows this variation as well as the optimum band gaps for the cells. The efficiencies quoted are for sunlight concentration of 1000 suns (AM1). The ideal values are plotted in Fig. 11.9(a). It can be seen that this approach ideally doubles the limit efficiency compared to a single-cell system. In actuality, such a system will create additional unavoidable optical losses compared to a single-cell system. Including optimistic values for these losses reduces the efficiencies to the more conservative values shown in Fig. 11.9(b). The advantage of multiple cells is reduced to 20 to 50%.

Considering a two-cell system, silicon is not an optimum choice for the low-band-gap cell. However, a silicon cell (band gap

Table 11.1. OPTIMUM BAND GAPS AND EFFICIENCIES FOR MULTIGAP CELLS (1000 × AM1)

Number of cells	System efficiency (%)	Band gaps (eV)									
1	32.4	1.4									
2	44.3	1.0	1.8								
3	50.3	1.0	1.6	2.2							
4	53.9	0.8	1.4	1.8	2.2						
5	56.3	0.6	1.0	1.4	1.8	2.2					
6	58.5	0.6	1.0	1.4	1.8	2.0	2.2				
7	59.6	0.6	1.0	1.4	1.8	2.0	2.2	2.6			
8	60.6	0.6	1.0	1.4	1.6	1.8	2.0	2.2	2.6		
9	61.3	0.6	0.8	1.0	1.4	1.6	1.8	2.0	2.2	2.6	
10	61.6	0.6	0.8	1.0	1.4	1.6	1.8	2.0	2.2	2.4	2.6
11	61.8	0.6	0.8	1.0	1.2	1.4	1.6	1.8	2.0	2.2	2.4	2.6

Source: After Ref. 11.10.

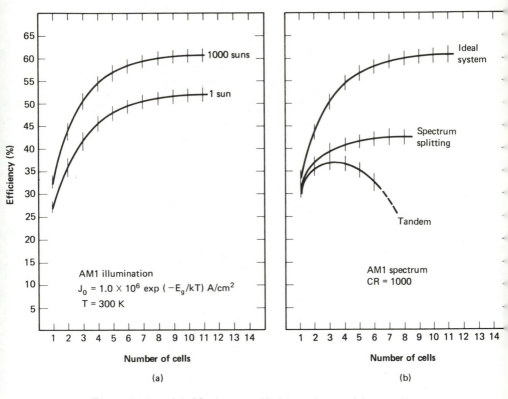

Figure 11.9. (a) Maximum efficiency for multigap cells for concentrated and unconcentrated sunlight. (b) Effect of optical losses.
(After Ref. 11.10, © 1978 IEEE.)

of 1.1 eV) combined with a material with a band gap in the region 1.6 to 2.1 eV gives close to the optimum performance (Ref. 11.11). The first really striking experimental results for the multigap approach were obtained in 1978 for a system consisting of a silicon cell and a $Al_xGa_{1-x}As$ heterojunction cell (Ref. 11.11). The band gap on the lower-band-gap side of the heterojunction was 1.61 eV. A wavelength-selective mirror was used to reflect photons of energy less than 1.65 eV onto the silicon cell, with the remainder passing through to the heterojunction cell. Under 165 suns concentration, the combined output of the system corresponded to an efficiency of 28.5%, the highest efficiency obtained for any photovoltaic system at the time.

  With the multigap approach, the voltage output of each band-gap cell and generally the current output will be different. It is of course possible to have an individual circuit for each cell type at the

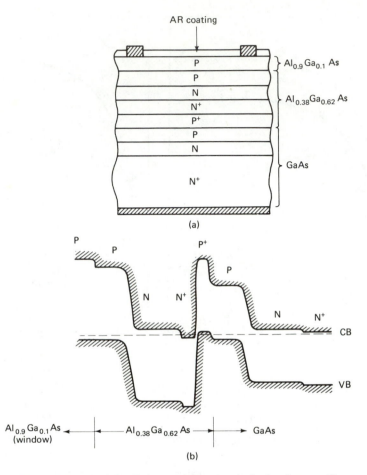

Figure 11.10. (a) Two series-connected tandem cells formed by epitaxial processes on the one substrate. (b) Corresponding energy-band diagram.

expense of increased complexity. A second alternative is to connect the cells in series. As mentioned in Section 6.6.4, the current output of series-connected cells is that of the worst cell. To maintain the efficiency of the multicell configuration, it is necessary to design the different types of cells so that each has the same short-circuit current. It has been suggested that this is also a reasonable criterion for selecting cell band-gaps to obtain maximum efficiency (Ref. 11.11).

An interesting idea in this area is the construction of tandem cells series interconnected on the one substrate. This can be accomplished using the same type of epitaxial growth technique as used to produce GaAs-based cells, as described in Section 10.4. For example, Fig. 11.10 shows the device structure and energy-band diagram of a solar cell structure that corresponds to two series-connected tandem cells (Ref. 11.12).

The top layer acts as a window layer for the underlying $Al_{0.38} Ga_{0.62} As$ cell. Under this cell are two heavily doped layers having several functions. These act as a back surface field for the top cell and a front surface field for the lower cell. The width of the depletion region at their junction is very thin and electrons can flow between the conduction and valence band at this junction by quantum mechanical tunneling process. This region consequently acts as the series connection between the cells, as well as an optical window for the underlying GaAs. Structures of this complexity are well within the bounds of the technology developed for fabricating semiconductor injection lasers (Ref. 11.13).

An important question for series-connected multigap cells is whether there are large variations in the relative spectral content of sunlight during the normal operating conditions of the cells. Such variations will cause changes in the relative values of the current output of the cell and have a pronounced effect on system efficiency. Preliminary data on this point indicates that, although variations can occur because of clouds and near sunset, this is not expected to be a major loss mechanism (Ref. 11.11).

11.6.3 Thermophotovoltaic Conversion

One important loss in solar cells is due to the fact that photons with energy far in excess of the band gap create only one electron–hole pair. Their effect on the cell output is consequently the same as a photon of much lower energy. Figure 11.11(a) shows

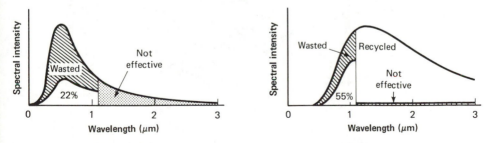

Figure 11.11. Utilization of the energy incident upon a silicon solar cell when illuminated by black body radiation at two different temperatures:
(a)   6000°C.   (b)   2000°C.
(After Ref. 11.15.)

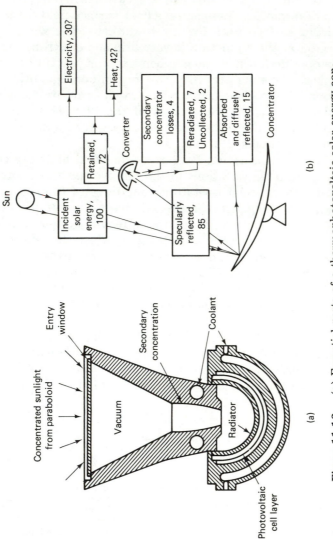

Figure 11.12. (a) Essential parts of a thermophotovoltaic solar energy converter. (b) Corresponding energy-conversion system and possible energy budget. (After Ref. 11.15.)

the fate of the incident energy as a function of wavelength for a silicon cell.

If a solar cell is illuminated by a black body at lower temperature (2000°C), the situation is modified as shown in Fig. 11.11(b). More of the energy would be utilized in those photons that are more energetic than the band gap. The efficiency of the solar cell would actually decrease, because relatively fewer photons would have energy larger than the band gap. However, if most of these ineffective photons could be redirected to the black body, thus contributing to maintaining its temperature by being absorbed in it, the situation changes. Such photons are no longer useless, but supply part of the energy required to hold the black body at high temperature.

In thermophotovoltaic solar energy conversion (Ref. 11.14), the sun is used to heat a radiator to high temperatures; the radiator then reemits radiation onto a solar cell. The long-wavelength radiation the cell cannot use is recycled to the radiator. The essential parts of a thermovoltaic converter are shown in Fig. 11.12. The rear of the cell is made highly reflective so that long-wavelength radiation passing through it is reflected back to the radiator. Although the theoretical upper limit of efficiency is very high for this concept, the fact that it involves several processes in series ensures that experimental efficiencies will be more modest (Ref. 11.15).

11.7 SUMMARY

Concentration of sunlight displaces the cost of a photovoltaic system from the cells to the cost of the concentrating and tracking elements. Continuous tracking of the sun is required to maintain concentration ratios greater than about 12.

Upper limits to cell efficiency at fixed temperature increase with increasing concentration ratio. This beneficial effect is offset by the generally higher cell operating temperatures when used in concentrating systems. For low concentration ratios, passive cooling of the cells is workable. For concentration ratios greater than about 50, active cooling such as by circulating water is required. This gives rise to total energy systems that produce both electricity and thermal energy from sunlight.

In concentrating systems, cell efficiency can be a more critical parameter than cell cost. This allows relatively complex schemes for increasing photovoltaic efficiency to be implemented. Concepts based on using several different band-gap cells to convert different parts of the sun's spectrum are likely to result in system efficiencies above 30%. Modifying the spectral content of sunlight as in the thermophotovoltaic effect could also result in similar efficiencies.

EXERCISES

11.1. Consider the case of a luminescent concentrator of Fig. 11.3. If the light reemitted by luminescence is emitted with uniform intensity in all directions, calculate the percentage trapped within the sheet by total internal reflection for the case where the point of emission is midway through the sheet. Assume that the sheet refractive index is 1.5.

11.2. A silicon shallow junction solar cell has a top-layer sheet resistivity of 30 $\Omega/\square$. At 1 sun, it gives its maximum power output at 450 mV and a current density of 30 mA/cm^2. Estimate the maximum spacing permissible between the fingers of the top contact to the cell to give a power loss due to lateral current flow in the top layer of less than 4% at 100 suns' operation.

11.3. At 1 sun operation (100 mW/cm^2), a solar cell gives an open-circuit voltage of 0.60 V and a short-circuit current of 0.6 A at 300 K. The design area of the cell is 20 cm^2, its ideality factor is 1.2, and its series resistance is 0.007 Ω. Assuming that the latter two parameters do not vary with light intensity, calculate and sketch the expected efficiency of this cell at 300 K as a function of concentration ratio over the range 1 to 50. [Use Eq. (5.17) to calculate the effect of series resistance on solar cell output.]

11.4. By referring to Fig. 5.1(b), select the ideal band gap under AM1.5 illumination for the second cell in a two-cell series connected tandem arrangement if the other cell is (a) silicon (E_g = 1.1 eV); (b) gallium arsenide (E_g = 1.4 eV). Calculate the limiting efficiency in each case under AM1.5 (83.2 mW/cm^2) and 1000 $\times$ AM1.5 irradiation if the cell temperatures are held at 300 K.

REFERENCES

[11.1] W. T. WELFORD AND R. WINSTON, *The Optics of Nonimaging Concentrators* (New York: Academic Press, 1978).

[11.2] A. RABL, "Comparison of Solar Concentrators," *Solar Energy 18* (1976), 93–112.

[11.3] J. L. WATKINS AND D. A. PRITCHARD, "Real-Time Environmental and Performance Testing of Concentrating Photovoltaic Arrays," *Conference Record, 13th IEEE Photovoltaic Specialists Conference, Washington, D.C.*, 1978, pp. 53–59;

M. W. EDENBURN, D. G. SCHUELER, AND E. C. BOES, "Status of the DOE Photovoltaic Concentrator Technology Development Project," *Conference Record, 13th IEEE Photovoltaic Specialists Conference, Washington, D.C.*, 1978, pp. 1028-1039.

[11.4] D. R. MILLS AND J. E. GIUTRONICH, "Ideal Prism Solar Concentrators," *Solar Energy 21* (1978), 423-430.

[11.5] C. F. RAPP AND N. L. BOLING, "Luminescent Solar Concentrators," *Conference Record, 13th IEEE Photovoltaic Specialists Conference, Washington, D.C.*, 1978, pp. 690-693.

[11.6] J. A. CASTLE, "10 kW Photovoltaic Concentrator System Design," *Conference Record, 13th IEEE Photovoltaic Specialists Conference, Washington, D.C.*, 1978, pp. 1131-1138.

[11.7] R. L. DONOVAN et al., "Ten Kilowatt Photovoltaic Concentrating Array," *Conference Record, 13th IEEE Photovoltaic Specialists Conference, Washington, D.C.*, 1978, pp. 1125-1130.

[11.8] E. C. BOES, "Photovoltaic Concentrators," *Conference Record, 14th IEEE Photovoltaic Specialists Conference, San Diego*, 1980, pp. 994-1003.

[11.9] R. SAHAI, D. D. EDWALL, AND J. S. HARRIS, JR., "High Efficiency AlGaAs/GaAs Concentrator Solar Cell Development," *Conference Record, 13th IEEE Photovoltaic Specialists Conference, Washington, D.C.*, 1978, pp. 946-952.

[11.10] A. BENNETT AND L. C. OLSEN, "Analysis of Multiple-Cell Concentrator/Photovoltaic System," *Conference Record, 13th IEEE Photovoltaic Specialists Conference, Washington, D.C.*, 1978, pp. 868-873.

[11.11] R. C. MOON et al., "Multigap Solar Cell Requirements and the Performance of AlGaAs and Si Cells in Concentrated Sunlight," *Conference Record, 13th IEEE Photovoltaic Specialists Conference, Washington, D.C.*, 1978, pp. 859-867.

[11.12] S. M. BEDAIR, S. B. PHATAK, AND J. R. HAUSER, "Material and Device Considerations for Cascade Solar Cells," *IEEE Transactions on Electron Devices ED-27* (1980), 822-831.

[11.13] E. W. WILLIAMS AND R. HALL, *Luminescence and the Light Emitting Diode*, Vol. 13, International Series on Science of the Solid-State, ed. C. R. Panydin (Oxford: Pergamon Press, 1978).

[11.14] R. M. SWANSON, "A Proposed Thermophotovoltaic Solar Energy Conversion System," *Proceedings of the IEEE 67* (1979), 446-447.

[11.15] R. N. BRACEWELL AND R. M. SWANSON, *Proceedings of the Electrical Energy Conference*, Institute of Engineers, Australia, Publication 78/3, May 1978, pp. 52-55.

Chapter

12

PHOTOVOLTAIC SYSTEMS: COMPONENTS AND APPLICATIONS

12.1 INTRODUCTION

In previous chapters, the properties of the most important part of any photovoltaic system, the solar cells themselves, have been treated in some detail. In this and the following chapters, the other components required for a photovoltaic system will be described, as will the performance and commercial viability of overall systems.

Since the output of solar systems is intermittent and unpredictable in the terrestrial environment, some form of energy storage and/or auxiliary energy supply is required if the system is to supply power on demand. The range of available and potential storage options will be explored.

Solar cells generate a dc (direct-current) output. The voltage at which the maximum power can be extracted changes with sunlight intensity and cell temperature. Since electrical power is most commonly utilized in ac (alternating-current) form, some form of power conditioning is required between the solar cell modules and the elec-

222

trical load. The general features desirable for power conditioning equipment are also outlined.

In the past the costs of solar cells have restricted their commercial use to relatively small stand-alone power supplies in remote regions. With the continuing reduction in cell prices, the range of commercially viable applications expands (Ref. 12.1). Some of these are described in the following sections.

12.2 ENERGY STORAGE

12.2.1 Electrochemical Batteries

Electrochemical batteries have been the predominant choice for storage for photovoltaic systems installed in the past. Lead–acid batteries and, to a lesser extent, nickel–cadmium batteries have been used. The major difficulty with this form of storage is its high cost together with the vast amount of material required to implement it on a large scale.

Several alternative battery systems are under development for potential use in electric vehicles and for short-term storage ("load leveling") in electricity supply systems (Ref. 12.2). These hold the promise of substantially lower battery costs for photovoltaic applications in the future (Ref. 12.3). The most promising systems at the present point include the zinc–chlorine and high-temperature batteries such as the sodium–sulfur and lithium–iron sulfide.

One development in the area of electrochemical storage that seems particularly well suited to stand-alone photovoltaic systems is the redox battery (Ref. 12.4). The concept of a *redox couple* was introduced in Section 9.7.2. The term refers to oxidized and reduced states of a species within solution. In a redox battery, two separate redox couple solutions are kept largely isolated from each other. During charging, one couple is oxidized while the couple in the other solution is reduced. The opposite occurs during discharge.

The redox-couple solutions that have received the most attention are acidified chloride solutions of chromium (redox couple: Cr^{2+}/Cr^{3+}) and iron (redox couple: Fe^{2+}/Fe^{3+}). Figure 12.1 shows how these are arranged to form a redox system in its simplest form. Each tank contains one of the redox-couple solutions. These solutions are pumped through the power-conversion section, where they are kept separated by a highly selective ion-exchange membrane. Electrical contact is made to each solution by an inert carbon elec-

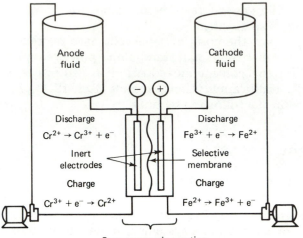

Figure 12.1. Schematic diagram of an electrically rechargeable redox energy storage system. (After Ref. 12.4.)

trode. The membrane prohibits the passage of iron and chromium ions but allows easy passage of chloride and hydrogen ions.

When the storage system is charged, the chromium solution contains chromium ions mostly in the reduced (Cr^{2+}) state, while the iron solution has its ions in the oxidized (Fe^{3+}) state. In the discharge mode, the following reactions occur:

1. At the anode the chromium ions are oxidized:

$$Cr^{2+} \longrightarrow Cr^{3+} + e^- \qquad (12.1)$$

2. At the cathode, the iron ions are reduced:

$$Fe^{3+} + e^- \longrightarrow Fe^{2+} \qquad (12.2)$$

3. Across the membrane, H^+ ions move from anode to cathode and Cl^- ions more in the opposite direction to maintain neutrality.

Electrons flow in the external circuit from the anode to the cathode, allowing current to be extracted from the cell terminals. A voltage impressed across the cell terminals encourages the reaction to proceed in the opposite direction, charging the cell. A number of these cells can be connected in parallel hydraulically and in series electrically to increase the voltage output.

224

A feature of the redox system that differs from common batteries is the independent sizing of the system power by the size of the power-conversion section and of the storage capacity by the selection of tank volumes and solution concentrations. This feature makes the system ideal for stand-alone photovoltaic systems, where storage of 1 week or more may be required to cover periods of low sunshine. The relatively mild solutions permit the use of inexpensive plastic materials for storage tanks and pipes. There is also no fundamental limitation to the number of charge/discharge cycles the system can tolerate, and operating lives up to 30 years have been predicted. The disadvantage of the approach lies in the relatively low power density of the electrolytes. A given volume of charged solutions can produce the same amount of electricity as about $\frac{1}{100}$ the volume of petroleum-based fuels, although a major difference is that the solutions can be recharged.

12.2.2 Large-Capacity Approaches

Batteries are an energy storage medium suited for both very small and very large photovoltaic systems. As discussed in Chapter 14, energy storage as an element of conventional electricity supply networks acts to improve the viability of including solar electric generation into the system. In this regard, it is worth noting that several large-capacity energy storage techniques are already being incorporated into such networks.

The most established technique for storing electrical energy on a large scale is *pumped hydro storage.* During periods of low demand, energy is stored by pumping water from a reservoir at a low level to a reservoir at a higher level. During periods of high demand, water is allowed to flow in the opposite direction to drive turbines, generating power. About two-thirds of the original electrical energy can be recovered in this approach. This method is presently limited by the shortage of suitable sites. A proposed development of this system in which the lower reservoir is constructed several hundred meters underground in hard rock eliminates this disadvantage to some extent. The larger "heads" obtainable also reduce the size of the upper reservoir for a given storage capacity (Ref. 12.2).

In a *compressed-air* energy storage plant, excess energy is used to store compressed air in an underground reservoir. Although this technique is more complex in practice than the pumped hydro method, it does have the advantages of a higher density of energy storage and greater flexibility in siting the underground reservoir

(Ref. 12.2).   Installations can be smaller and still be economically viable.   The world's first commercial installation at Huntorf in West Germany has an operating capacity of over $\frac{1}{2}$ million kilowatt-hours. Pumped hydro storage has to be an order of magnitude larger to obtain the full economic benefit.

Storage of electrical energy by conversion to hydrogen is another possible route well suited to photovoltaics because of the low-voltage dc requirements of electrolysis.   In fact, as seen in Section 9.7.3, photoelectrolysis can be achieved directly at the surface of a semiconductor, although present efficiencies are low.   Hydrogen has several advantages as an energy storage medium.   It can be economically transferred over large distances by pipeline.   It is suitable for use as a fuel in conventional engines for motive power or in a fuel cell to generate electricity efficiently.   These features lead to concepts of a *hydrogen economy*, where hydrogen forms the basic fuel for mankind (Ref. 12.5).   From the point of view of storage for photovoltaics, a major disadvantage involves the present low storage efficiencies ($<50\%$) attainable.

Storage in superconducting magnets or as mechanical energy in flywheels are other possibilities, although both seem intrinsically more expensive than other alternatives at the present time.

12.3   POWER CONDITIONING EQUIPMENT

A photovoltaic system in general may consist of the cells, a storage medium, some form of backup either in the form of an auxiliary generator or the electricity supply grid, and electrical loads, either ac or dc.   Power conditioning and control are required to provide an interface between these different system elements.   This is illustrated schematically in Fig. 12.2.

The simplest solar cell system is one in which the cells can be connected directly to the load, supplying power whenever there is adequate illumination.   Water pumping using a dc motor to drive the pump is an example of such a system.   The next simplest systems are those in which a dc load is to be supplied in a situation where sufficient battery storage can be installed so that no backup generator is required.   In this case, all that may be required is a *regulator* to prevent damage to the batteries by overcharging in sunny periods.   At the next level of complexity, a similar system is used to provide power to an ac load.   In this case, an *inverter* is required to convert the dc output of the solar cells and storage batteries to alternating form.   At a higher level of complexity, a backup generator (which

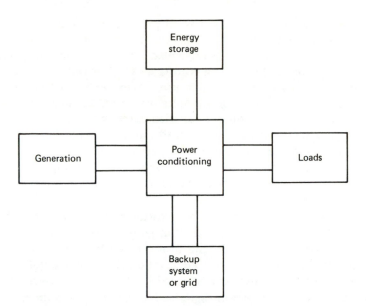

Figure 12.2. Power conditioning equipment in the most general case serves not only as an interface between the different components of a photovoltaic system but also performs control and protection functions.

may be the electricity supply grid) is also involved. In this case, some form of control is required to determine when the auxiliary supply is operated.

The major developmental effort in the area of power conditioning is in improving the performance and lowering the cost of inverters. Key performance parameters are efficiency and no load power drain (Ref. 12.6).

12.4 PHOTOVOLTAIC APPLICATIONS

In the past, the high costs of solar cells restricted their commercial use to applications where small amounts of power were required in regions remote from an electricity supply grid. Telecommunication systems formed the backbone of this commercial market. These have ranged from microwave repeater stations requiring several peak kilowatts of generating capacity per station to small modules rated at only tens of watts for remote-area radio telephone services.

Other reasonably large uses have been in powering navigational aids and warning devices, railway crossings, weather and pollution monitoring equipment, corrosion protection using the impressed current technique, as well as in powering consumer products such as calculators and watches. Cells have also been used for powering educational television in developing countries and for providing refrigeration in conjunction with immunization programs in such countries.

As cell costs continue to decrease, other applications particularly relevant to developing countries have become economically viable (Ref. 12.7). Examples are water pumping for irrigation on a small scale and water purification to provide drinking water. Developmental aid programs may be one way of supplying cells to this market which would overcome likely problems with raising capital for such systems.

The first application of photovoltaics likely to have a measurable impact on world energy demands is in supplying residential power in North America. In the mode of operation envisaged, the residence would also be connected to the grid which would act as a long-term storage medium, as discussed in Chapter 14. Several of the technologies based on silicon discussed in Chapter 7 appear capable of producing cells at a cost consistent with this use.

For larger-scale generation of power, such as in a central power plant of large capacity, cells have to be about half the cost as in this residential use to be competitive. A technology based on thin-film solar cells has a much higher probability of reaching these costs. Other desirable features for this mode of operation are discussed in Chapter 14.

12.5   SUMMARY

In any photovoltaic system, components other than the solar cells alone are required in all but a few applications. A photovoltaic system may involve the cells, energy storage, power conditioning and control equipment and a backup generator. The major item of power conditioning equipment is generally an inverter to convert the dc output of cells and battery storage to the alternating form often required by the load.

In the past, the commercial use of solar cells has been restricted to supplying relatively small amounts of power in remote areas. In the future, a much wider range of applications will become viable as cell costs continue to decrease. The supply of residential power in grid-connected areas of the United States is seen as a potentially large application that is economically workable with technology presently at an advanced stage of development. To be usable for the generation of power in a large-capacity central power plant, cell costs need to be about half that required for the residential application. A thin-film technology that uses minimal semiconductor material is the approach most likely to produce cells at such costs.

EXERCISE

12.1. Battery storage is required for a photovoltaic system that has to supply a peak load of 10 kW and an average load of 1 kW. Assume that an advanced lead-acid battery capable of supplying peak load costs $100 per kWh of storage, whereas a redox battery costs $300 per kW peak rating of the power conversion section plus $40 per kWh of energy storage. Which system would cost less to purchase for: (a) 4 h of storage; (b) 5 days of storage?

REFERENCES

[12.1] D. COSTELLO AND D. POSNER, "An Overview of Photovoltaic Market Research," *Solar Cells 1* (1979), 37–53.

[12.2] F. R. KALHAMMER, "Energy Storage Systems," *Scientific American 241*, No. 6 (December 1979), 42–51.

[12.3] *Handbook for Battery Energy Storage in Photovoltaic Power Systems*, Final Report, DOE Contract No. DE-AC03-78ET 26902, November 1979.

[12.4] L. H. THALLER, "Redox Flow Cell Energy Storage Systems," Report No. DOE/NASA/1002-79/3, NASA TM-79143, June 1979.

[12.5] J. O'M. BOCKRIS, *Energy: The Solar Hydrogen Alternative* (London: Architectural Press, 1975).

[12.6] G. J. NAAIJER, "Transformerless Inverter Cuts Photovoltaic System Losses," *Electronics 53*, No. 18 (August 14, 1980), 121–126.

[12.7] L. ROSENBLUM et al., "Photovoltaic Power Systems for Rural Areas of Developing Countries," *Solar Cells 1* (1979), 65–79.

Chapter 13

DESIGN
OF STAND-ALONE SYSTEMS

13.1 INTRODUCTION

The major commercial market for photovoltaic systems in the past has been for small, reliable power supplies in remote areas. These generally operate without any nonsolar backup and are therefore the sole source of electric power for the load. The design of such systems is discussed in this chapter.

A schematic of a simple solar-powered system is shown in Fig. 13.1. The load for most such small systems utilizes power in the dc form as generated by the solar cells. In addition to the array of solar cell modules and storage batteries, the other system components are a blocking diode to prevent loss of battery charge through the cells overnight and a regulator to prevent overcharge of the battery in periods of high insolation levels.

13.2 SOLAR MODULE PERFORMANCE

Present solar cell modules generally contain sufficient series-connected solar cells to generate enough voltage to charge a 12-V

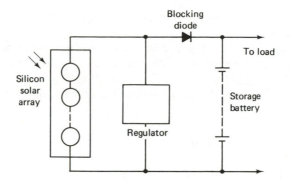

Figure 13.1. Simplified stand-alone solar power system. (After Ref. 13.2.)

battery. Modules are connected in series to increase the system voltage output and in parallel to increase the output current of the system. The number of series-connected cells required to charge a nominally 12-V system is higher than first expected for a number of reasons. For lead–acid batteries, over 14 V is required to fully charge a nominally 12-V battery. If a silicon blocking diode is used, a further 0.6 V at the minimum is required to forward-bias it. Additionally, module temperature frequently will exceed 60°C in the field. The open-circuit voltage of the module decreases by about 0.4% for each 1°C increase in temperature (Section 5.3). This means that there will be about a 3-V reduction in open-circuit voltage for a module with a 20-V open-circuit voltage specified at 25°C. As indicated in Section 6.6.2, different module designs will cause cells to reach different temperatures in the field. Modules mounted to allow air circulation at the rear will run cooler than those without this feature.

For best performance, modules are mounted so that they face south in the northern hemisphere and north in the southern hemisphere at an angle from the horizontal which depends on the latitude of the site. For maximum output over the year, this angle approximately equals the latitude angle. For systems as described in this chapter with 10 to 30 days of battery storage, the optimum value of this angle increases by about 15° to boost the winter output of the system.

A module designed to charge a 12-V battery will generally be able to generate sufficient voltage during daylight hours to do so. The current output will be very closely proportional to the intensity of sunlight on the module. Hence, the focus of attention in the

231

design of the systems of the present chapter is on the current output of the module.

A final consideration regarding module performance is the effect of accumulated dust. This will be a cyclic effect with minimum blocking occurring after rain. For glass-covered modules, available data indicate that the loss due to this effect averages 5 to 10%.

13.3 BATTERY PERFORMANCE

13.3.1 Performance Requirements

The feature of photovoltaic systems that makes them competitive at present prices is high reliability and low maintenance costs. To achieve these characteristics, systems are designed with large secondary battery stores to tide them over the worst conceivable insolation periods. Battery maintenance is, in principle, the major maintenance requirement of stand-alone photovoltaic systems.

With such large battery storage, the charge/recharge cycle imposed on the battery is a seasonal cycle in which the battery charges during the summer and discharges during winter. Superimposed upon this seasonal cycle is a much smaller daily cycle in which the battery charges during the day and loses a small percentage of its charge at night. Because of the seasonal nature of the storage, a battery with low self-discharge characteristics is essential. High charge storage efficiency (ratio of charge that can be extracted from battery to charge put into battery) is also desirable.

13.3.2 Lead–Acid Batteries

The most commonly used batteries for solar systems have been lead–acid batteries. Lead–antimonial types as commonly used in automobiles are not suitable for professional solar systems because of their high rate of self-discharge (up to 30% of capacity per month) and low life.

Commercial batteries most suited for stand-alone power supplies have been those intended for *stationary* or *float* service. These are designed for applications such as serving as emergency power sources in uninterruptible power supplies. In this use, the batteries are kept fully charged but immediately take up the load demand if the primary power source fails. Operating life in this type of application is generally in excess of 15 years. These batteries are normally rated at an 8- or 10-h discharge rate and have either lead–calcium or pure lead plates. More recently, this type of battery has been developed to meet the specific requirements of the photovoltaic mode of operation (Ref. 13.1).

In the type of solar system described in this chapter, the battery will operate in a fairly unusual mode whereby it remains fully charged in summer but only partially charged through most of the winter. Long periods in a low state of charge can cause much larger crystals of lead sulfate to form on the battery plates than the small crystals which normally form during discharge. This process, known as *sulfation*, leads to loss of capacity and reduced battery life. Good design practice is to ensure that the battery store is large enough so that it retains a reasonable percentage of full charge during the winter months. This will also ensure that the concentration of sulfuric acid in the electrolyte solution remains relatively high in these months, reducing the possibility of freezing (Ref. 13.1).

During summer, solar cells will be generating excess power over that required by the load and it would be possible to overcharge the battery. This is undesirable on several accounts. It can cause the evolution of hydrogen and oxygen from the battery, a process known as *gassing*, with associated loss of electrolyte and safety hazard. It can also lead to excessive plate growth and the shedding of active material from the plates, reducing battery life. On the other hand, good practice with lead–acid batteries is to allow for a periodic boost charge. The resultant gassing agitates the electrolyte, preventing *stratification* of more concentrated material in lower regions. An overcharge or *equalizing charge* also ensures that weaker cells in the battery have the opportunity to become fully charged (Ref. 13.1).

Figure 13.2 shows how the voltage across a battery cell varies with the amount of charge stored in it when charged with a constant

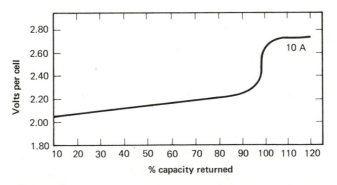

Figure 13.2. Constant-current charging characteristics of a lead–acid battery suitable for solar use. (After Ref. 13.2.)

current. At about 95% recharged, there is a sharp rise in voltage across the cell. This corresponds to the gassing point. To restrict the amount of gassing, at the same time providing for the beneficial effects of a periodic overcharge, a reasonable compromise is to use a voltage regulator to restrict the voltage per battery cell to about 2.35 V for the case shown (Ref. 13.2).

Other relevant considerations are the variation in battery capacity with rate of discharge and temperature. Battery capacity is usually specified at a given discharge rate. For example, Fig. 13.3 shows the measured ampere hours extracted from a battery at two different discharge rates. The battery capacity was specified as 550 A-h when discharged to 1.85 V per cell at a 10-h rate (55 A drawn continuously for 10 h). The battery capacity at the 10-h rate can be seen to be above specifications. At the 300-h rate that is more typical of solar operation, the battery capacity is nearly double the specified value. It follows that in designing a solar system, not only the battery capacity but also the discharge rate at which it is specified are important.

The storage capacity decreases with temperature, which is unfortunate because winter is when most use is made of storage. As a rule of thumb, capacity decreases by about 1% per °C below about 20°C. This, together with the possibility of freezing the electrolyte, makes it desirable to insulate the battery from abnormally cold environments. At the other extreme, high temperatures accelerate battery aging mechanisms, increase the rate of self-discharge, and increase the use of electrolyte. Again, the batteries require suitable housing to avoid high temperatures.

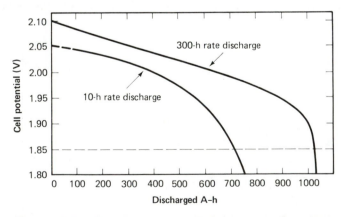

Figure 13.3.   Constant-current discharge curves for a lead–acid battery at different discharge rates. (After Ref. 13.2.)

At moderate charge and discharge rates, about 80 to 85% of the charge put into a lead-acid battery can be recovered upon discharge. However, a large part of this inefficiency arises from gassing which occurs during charging. In the stand-alone photovoltaic mode, gassing is unlikely to occur during winter, when an appreciable amount of the load is being supplied from the battery. Hence, the charge storage efficiency in critical months will be much higher than the previous figure. Coulombic efficiencies as high as 95% have been quoted (Ref. 13.3).

13.3.3 Nickel-Cadmium Batteries

Nickel-cadmium batteries of the pocket-plate type have also been used in solar systems. Their major advantages compared to the lead-acid batteries described in Section 13.3.2 are:

1. Ability to be overcharged without damage
2. Ability to spend long periods only slightly charged without damage
3. Mechanically more rugged, making them more transportable
4. Ability to withstand freezing without damage

Their major disadvantages are:

1. Higher cost (about three times higher in large volume for a given storage capacity)
2. Low charge storage efficiency (55 to 60% for solar operation)
3. Significantly less capacity increase due to the low discharge rate in solar applications

At the present point in time, their advantages do not outweigh their disadvantages in most solar applications.

13.4 POWER CONTROL

A blocking diode is normally inserted between the battery and solar array to prevent the battery losing charge through the array at night. The voltage drop across the diode subtracts from the array voltage when it is supplying power to the battery. For silicon diodes, this

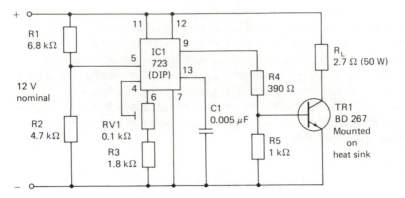

Figure 13.4. Shunt regulator for a 12-V/60-W solar array. (After Ref. 13.2.)

will be about 0.6 to 0.9 V but can be reduced to 0.3 V by the use of Schottky or germanium diodes.

To protect batteries from overcharging, some form of voltage regulation is required. For small systems, a simple linear shunt regulator can be used to dissipate the unwanted power. One possible design (Ref. 13.2) is shown in Fig. 13.4 for a 12-V system of up to 60 W of generating capacity. RV1 is adjusted to set the level at which the regulator cuts in. About 14.1 V is recommended. Any attempt to charge the battery voltage above this level will result in the charging current being shunted through R_L and TR1.

For large arrays, this technique is not feasible because of the large quantities of heat generated. A preferable method is to dissipate the extra energy as heat within the distributed solar cells. This can be accomplished by either short-circuiting or open-circuiting a portion of the solar array.

A schematic of a short-circuiting type of regulator is shown in Fig. 13.5. Transistors successively short out parallel sections of the array to maintain battery voltage below some desired limit. Although shorting an individual cell is completely acceptable, *problems can arise if a string of series-connected cells is shorted.* A cell generating below-average current can become reverse-biased and be made to dissipate virtually the entire peak module output, as discussed in Section 6.6.4. Field failures with this short-circuiting mode of regulation have been common. Consequently, it is not recommended unless special protective features such as bypass diodes have been incorporated into the modules that make up the array.

An alternative technique is to open-circuit the parallel sections. One approach using thyristors is shown in Fig. 13.6. Consid-

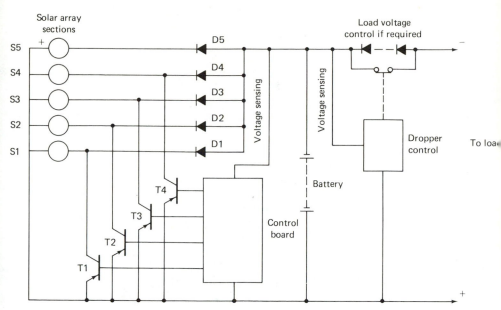

Figure 13.5. Possible technique for implementing a short-circuiting type of regulator for larger arrays. This type of regulation is not recommended unless protective features such as bypass diodes are incorporated into the array sections. (After Ref. 13.2.)

ering solar array group 1, pulses are directed to the base of either TR1 or TR4, depending on battery voltage. Pulses to TR1 ensure that TH1 is nonconducting, open-circuiting that section of the array. Pulses to TR4 switch the thyristor to its conducting state. In this state it fulfills the role of the normal blocking diode. Note that a degree of hysteresis is built into the voltage-sensing circuit to prevent instabilities. The battery voltage and array current obtained with such a system on a summer day when the battery is approaching a fully charged state is shown in Fig. 13.7.

13.5 SYSTEM SIZING

To size a solar system, it is essential to have accurate information about the load to be supplied and the best available radiation data for the intended site.

In cases such as microwave repeater stations, the load is easily determined since it is virtually constant. In other applications, such

237

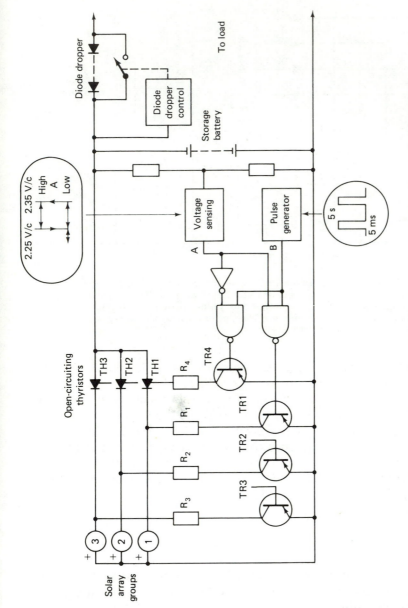

Figure 13.6. Thyristor open-circuiting mode of regulation. (After Ref. 13.2.)

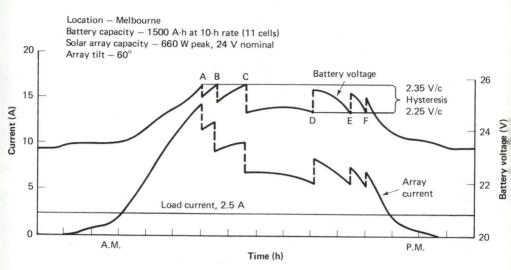

Figure 13.7. Operating characteristics of the method of regulation of Fig. 13.6. (After Ref. 13.2.)

as subscriber radio-telephone, the larger power demands in the transmitting mode makes the load dependent on the user call rate and hence highly unpredictable. Detailed radiation data are unlikely to be available for the type of remote locations where stand-alone solar systems are used. The best that can be done is to extrapolate from data recorded at stations in similar locations. A good source for data where more detailed information is not available is Ref. 13.4.

Solar cell suppliers and major users have developed computer programs for sizing photovoltaic systems. These programs can be quite sophisticated, taking into account such effects as the variation of solar and battery voltages upon temperature and capacity decrease in batteries with temperature. A simpler design technique will now be described which illustrates the concepts involved and which should be adequate where lack of data does not warrant a more detailed design.

The first step in this design procedure is to select the capacity required for the battery storage. The battery storage capacity can be regarded as serving two purposes. One is to provide a *reserve capacity* to cover an exceptionally long period without sunshine, or system failure. The other purpose is to provide seasonal storage (Ref. 13.3).

The amount of reserve capacity required depends on several factors. Climate is one. Sunny, arid regions require less than foggy coastal regions. The accessibility of the site, the regularity of system monitoring, and the consequences of system failure are other important considerations. Normally, reserve capacity in the range 10 to 20 days is adequate, with 30 days a conservative extreme. In selecting a battery to give this capacity, consideration should be given to the effect of temperature and discharge rate on battery capacity.

Having selected the reserve capacity, the next step is to decide upon the depth of discharge acceptable for the system battery under normal seasonal fluctuations in solar energy input. Excessive depth of discharge will shorten lead–acid battery life as indicated in Section 13.3.2. Discharge to a depth of 50% of available capacity is the maximum desirable. On the other hand, designing for very shallow discharge will increase the size of the solar array required. As costs of modules decrease, optimum designs shift to shallower discharge depths.

Once the depth of discharge due to seasonal fluctuations is selected, the total battery capacity can be calculated. Since the reserve capacity (C_R) should be available even when the battery is at its lowest state of charge due to seasonal variations, the total capacity required is $C_R/(1 - d)$, where d is the fractional depth of discharge desired.

The size of the battery is now fixed. The next step is to select the size of solar array required. Current and voltage outputs are sized separately. The voltage output is selected to be large enough to allow the battery to be charged efficiently throughout the year. The current output is chosen to ensure that the battery does not discharge below the selected depth of discharge due to seasonal effects.

To proceed further, radiation data are required. Such data will normally be in the form of global radiation on a horizontal surface (R), possibly together with the diffuse radiation on such a surface (D). If the latter is not available, it is possible to provide a reasonable estimate using the technique described in Ref. 13.5. To convert the available data to radiation falling on an array of other than horizontal orientation, some assumptions must be made. Assuming daily data, the mean direct daily radiation on a horizontal plane (S) is given by

$$S = R - D \tag{13.1}$$

From Fig. 13.8(a), the direct component on a surface at an angle β to the horizontal is given by

$$S_\beta = S \frac{\sin (\alpha + \beta)}{\sin \alpha} \tag{13.2}$$

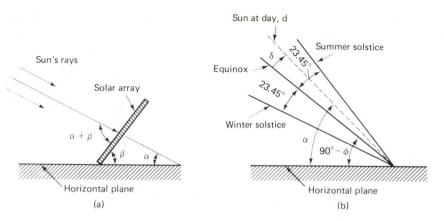

Figure 13.8. (a) Solar radiation at solar noon on a solar array tilted at angle β to the horizontal. Angle α is the altitude of the sun. (b) Relation between altitude of the sun (α), declination (δ), and latitude (φ).

where α is the noon altitude of the sun. From Fig. 13.8(b), α is given by

$$\alpha = 90° - \phi \pm \delta \qquad (13.3)$$

The plus sign applies for the northern hemisphere, the negative for the southern. Here ϕ is the latitude and δ the declination of the sun, given by

$$\delta = 23.45° \sin\left[\frac{360}{365}(d - 81)\right] \qquad (13.4)$$

where d is the number of days from the beginning of the year. Assuming that the diffuse radiation is independent of the array inclination, the total radiation on the array is given by (Ref. 13.2)

$$R_\beta = S\frac{\sin(\alpha + \beta)}{\sin\alpha} + D \qquad (13.5)$$

Equation (13.5) is strictly correct only at midday, but it provides a reasonable approximation for converting daily horizontal radiation to radiation on an inclined surface. More elaborate approaches would give more accurate results (Ref. 13.5).

The design procedure can best be described by working through a concrete example. We will design a solar system for Mel-

241

bourne, Australia (latitude 37.8°S), capable of supplying 100 W continuously to a load at 24 V dc.

For this installation, a storage *reserve capacity* of 15 days is selected. Since the load requires 100 A-h/day, this corresponds to 1500 A-h of storage. Selecting 25% as the design depth of discharge due to seasonal fluctuations in solar intensity to prolong battery life gives a total installed capacity of 2000 A-h [1500 A-h/(1 - 0.25)] at a discharge rate of 480 h (20 days).

The next stage in the design is to find the size of the solar module required to ensure that the batteries will not normally be discharged below 25% depth of discharge. The optimum angle of tilt of the fixed array will be the latitude angle plus 15 to 20°, say 60° for the present location. Once this angle to the horizontal, β, is selected, the raw radiation data are converted to radiation on a surface of this inclination. This is shown for Melbourne in Table 13.1. The average daily radiation on any array orientated 60° from the horizontal at this location is 21.0 MJ/m². A lower bound for the current-generating capability of the array can be estimated by equating the input charge to the array over the year to the charge required by the load. This would represent the situation for an ideal system with infinite storage. In the present case, the load requires 100 A-h/day (100 W ÷ 24 V × 24 h). The average daily radiation intensity on the inclined surface is 21.0 MJ/m², which, dividing by 3.6, converts to 5.83 kWh/m² or 583 mWh/cm². This corresponds to 5.83 h of bright sunshine at peak sunshine intensity of 100 mW/cm². Hence, the solar array peak current rating at 100 mW/cm² radiation intensity has to be at least 17.2 A (100 A-h/5.83 h).

Including the fact that not all charge supplied to the battery can be extracted from it but only an estimated 95% of charge for a battery below the gassing point, and allowing 10% degradation in performance for the average effect of dust accumulation, increases this estimate to 20.1 A (17.2 ÷ 0.95 ÷ 0.90).

An upper bound can be estimated by repeating this calculation using the radiation data for the worst month. If a system were designed in this way, the batteries would be very close to fully charged except during periods of inclement weather. For the present example the worst month is June, in which there is only the equivalent of 4.26 h of full sun radiation on the inclined surface. This gives an upper bound of 27.5 A (20.1 × 5.83 h ÷ 4.26 h).

The optimum array peak current rating will lie somewhere between these extremes. It can be found by a trial-and-error calculation, which involves checking on the battery state of charge throughout the year. For the present example, the optimum peak current

Table 13.1 Data for Stand-Alone Solar System Design

System Data

Site: Melbourne, latitude 37.8°S
Load: 100 W, 24 V
Array inclination: 60° to horizontal
Battery capacity: 2000 A-h
Peak array current rating: 25 A

Month	Average daily radiation (mWh/cm^2)			Monthly ampere hours			Battery state		
	Global (R)	Diffuse (D)	On array*	Array†	Load‡	Difference	Start	Finish	% full charge
Jan.	839	210	688	4559	3147	1412	2000	2000	100
Feb.	708	149	648	3878	2842	1036	2000	2000	100
Mar.	562	166	609	4035	3147	888	2000	2000	100
Apr.	436	127	575	3687	3045	642	2000	2000	100
May	297	98	463	3068	3147	-79	2000	1921	96
June	246	79	426	2732	3045	-313	1921	1608	80
July	277	82	462	3061	3147	-86	1608	1522	76
Aug.	374	120	520	3446	3147	299	1522	1821	91
Sept.	516	148	596	3822	3045	777	1821	2000	100
Oct.	697	197	673	4459	3147	1312	2000	2000	100
Nov.	732	241	627	4021	3045	976	2000	2000	100
Dec.	890	214	701	4645	3147	1498	2000	2000	100

*Calculated using Eq. (13.5).

†Calculated as: Peak current rating × days in month × average daily radiation × charge efficiency × dirt accumulation factor/100 mW/cm^2

‡Including 3% discharge of peak battery capacity per 30-day month.

Source: Adapted from Ref. 13.2.

rating of the array is 25 A. The generated ampere-hours, consumed ampere-hours and battery state of charge as a function of time throughout the year is shown in Table 13.1. Note that the battery remains above the design figure of 75% charge throughout the year. If it dropped below this, the array size would have to be increased. If the state of charge always remained well above this figure, a more economical design would be to use a smaller array size. Although the results of calculations are shown for all months of the year, only those between which the monthly radiation is less than average need be considered. The calculation does indicate the large amount of power being thrown away in summer with this design approach. Figure 13.9 shows the battery state of charge as a function of time through the year for different array current ratings. It vividly illustrates the effect of a small increase in array size. For example, a 4% increase in array size would keep the batteries in a higher state of charge throughout the year than would doubling the storage capacity.

To complete the system sizing, the array voltage must be specified. For the design described above to be strictly valid, the array must be capable of supplying peak load current at the highest temperature it is likely to reach during normal operation, even with the batteries approaching full charge ($\sim$2.35 V per cell). For the design example, the array must be capable of supplying 25 A at a voltage of 29 V (12 $\times$ 2.35 V + 0.8 V for blocking diode) at its as-

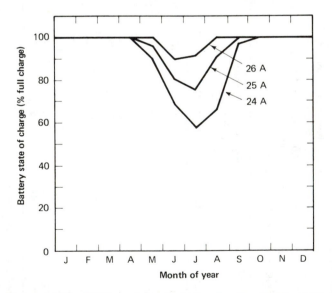

Figure 13.9. Battery state of charge for the design example in the text for three different array sizes.

sumed maximum temperature of 60°C. Therefore, the array must have a peak rating of 725 W at 60°C. Since the array power falls off at about 0.5% per °C, this corresponds to a peak array rating of 879 W at the normal specification temperature of 25°C and incident power level of 100 mW/cm^2.

However, this is a conservative design because of the large amounts of power that would be thrown away during summer. During the colder months, when the output of the module is most critical, weather data for Melbourne indicate that it would be unlikely for the ambient temperature to be above 20°C. For normal module designs, the cell temperature would be unlikely to exceed 44°C during these months. If an array output voltage of 29 V was specified at this temperature, the array would provide the calculated power during winter months. The output during summer would be less than calculated, although this may be no problem due to the large excess generated during these months.

The battery reaches its lowest state of charge in July, when the average temperature in Melbourne is about 10°C. Assuming that the battery operates at ambient temperature, this would require an increase in battery capacity to ensure that it still met the reserve-capacity requirement for this month. Assuming a 10% decrease in capacity at this temperature increases the required capacity to 2222 A-h at room temperature at a 480-h discharge rate. Both module and storage requirements may have to be modified to make them multiples of commercially available units. Sunnier regions of the world require smaller arrays to supply the same design load. For example, in some regions of Australia, arrays only two-thirds the size of the one calculated for Melbourne would supply this load (Ref. 13.2). A photograph of a solar power plant designed to supply loads of about this size is shown in Fig. 13.10. A shipping container is used to support the solar modules as well as to house the battery store, control electronics, and facilities for service crews.

To summarize the design procedure:

1. Determine load data.
2. Select battery size depending on latitude and any local climatic peculiarities.
3. Select an array inclination.
4. Estimate lower and upper bounds on array size from average and minimum monthly radiation falling on an array of the selected inclination.
5. Find the optimum array size to keep the batteries above a certain percentage charge throughout the year.

Figure 13.10. Solar power supply designed to provide approximately 100 to 150 W of electrical power continuously for a microwave repeater station. The shipping container supports the solar modules, as well as housing the storage batteries, control electronics, and facilities for servicing crews. (Photograph courtesy of Telecom Australia.)

6. Modify array inclination to find optimum if required.

7. Specify array voltage at the maximum operating temperature at which it is desired to supply full load current.

13.6 WATER PUMPING

Small-scale water pumping is an application well suited to the use of photovoltaics, for two reasons. First, solar arrays may be connected directly to the pump motor without requiring intermediate power conditioning or battery storage, giving a very simple, portable and potentially low maintenance system. Second, the requirement for pumped water decreases in periods of low solar insolation in many applications. This allows economical sizing of the system. Storage would be best implemented by storing pumped water.

One area where such small solar water-pumping units could make a big impact is in providing irrigation in less developed countries (Ref. 13.6). Irrigation can greatly improve the yield of a given area of land with consequent benefits. A photovoltaic powered microirrigation system ($\sim 250 W_p$) is well suited to the small plots cultivated by individual landholders. As it is unlikely that such landholders would be able to raise the necessary capital, an approach that has been advocated is to include such units in developmental aid programs (Ref. 13.6). This would serve two purposes. One would be to increase food supplies in these countries. The other would be to provide a substantial market for cells in the near term which would encourage their accelerated development.

13.7 SUMMARY

In principle, batteries produce the major maintenance requirements of stand-alone photovoltaic systems, requiring an annual or semi-annual topping up of electrolyte. However, the system must be carefully designed to produce maximum battery life. Overcharging of lead-acid batteries must be prevented, as must undercharging, which leaves the batteries with low stored charge over long periods.

Continuing decreases in the power demands of electronic communications equipment and in the costs of solar cells have made telecommunications the first major commercial terrestrial application for these cells. Stand-alone systems similar to those described in this chapter have been deployed in applications including microwave repeater stations, navigational aids, meteorological stations, and corrosion protection.

In the design of stand-alone systems, a large storage capacity is used to provide high reliability. The solar array is sized to ensure that this battery retains a reasonable percentage of full charge over the lower insolation winter months. In this design mode, the solar system does not produce the maximum possible power output over the year. For the most ideal stand-alone installations, the peak rating of the array must be about five times the average power the system is to provide. In marginal locations, this can increase by a factor of 2 or more.

EXERCISES

13.1. For the design example of Table 13.1, compare the typical daily cyclic variation in battery state of charge with the seasonal variation.

13.2. For a location at latitude $34°N$, find the angle of solar arry tilt that will maximize the system output for November, using the approximate method outlined in the text. The average global radiation at this location on a horizontal surface in November is $12 \, MJ/m^2/day$, and the corresponding figure for diffuse radiation is $4.1 \, MJ/m^2/day$.

13.3. Design a stand-alone photovoltaic system for a location at latitude $23°N$. The system is to supply a constant load of 250 W at 48 V dc. Starting at January, the global figures for radiation on a horizontal surface for the 12 months of the year are (the numbers in parentheses are the corresponding figures for diffuse radiation): 15.5 (3.2), 17.2 (4.2), 21.6 (4.0), 23.3 (6.0), 24.9 (7.0), 24.1 (8.8), 23.8 (8.9), 22.9 (8.1), 20.7 (7.3), 18.9 (4.8), 15.6 (4.7), and 14.5 (3.8) $MJ/m^2/day$, respectively.

REFERENCES

[13.1] *Handbook for Battery Energy Storage in Photovoltaic Power Systems*, Final Report, DOE Contract No. DE-AC03-78ET 26902, November 1979.

[13.2] M. MACK, "Solar Power for Telecommunications," *Telecommunication Journal of Australia 29*, No. 1 (1979), 20–44.

[13.3] *Solar Electric Generator Systems: Principles of Operation and Design Concepts*, booklet prepared by Solar Power Corporation.

[13.4] G. O. G. LÖF, J. A. DUFFIE, AND C. O. SMITH, *World Distribution of Solar Radiation*, Solar Energy Laboratory, University of Wisconsin, Report No. 21, July 1966.

[13.5] S. A. KLEIN, "Calculation of Monthly Average Insolation on Tilted Surfaces," *Solar Energy 19* (1977), 325–329.

[13.6] D. V. SMITH AND S. V. ALLISON, *Micro Irrigation with Photovoltaics*, MIT Energy Laboratory Report, MIT-EL-78-006, April 1978.

Chapter

14

RESIDENTIAL AND CENTRALIZED PHOTOVOLTAIC POWER SYSTEMS

14.1 INTRODUCTION

In this, the final chapter of this book, issues surrounding potential long-term applications of solar cells are raised. Two areas where photovoltaics could produce significant contributions to world energy demands is in supplying residential power and in generating electricity in large, centralized power plants. Because of the massive scaling up of cell production that generation at this level would entail and the finite diffusion rates of a new technology into commercial use, it is unlikely that this contribution will be more than a few percent prior to the turn of the century. This does not mean that there will not be systems of this type operating economically well before this date.

In large-scale production, economic analyses indicate that most of the advanced silicon technologies described in Chapter 7 can produce solar cell modules which can be sold at a cost that makes them competitive for supplying residential power. The demands placed on module costs for photovoltaics to compete in the form of large, centralized power plants are about twice as severe. As discussed in Chapter 10, thin-film photovoltaic devices have the best prospects for attaining the ultimate low in module costs.

14.2 RESIDENTIAL SYSTEMS

14.2.1 Storage Options

In Chapter 13, the use of photovoltaics was described in stand-alone configurations for remote areas. However attractive it might seem to be independent of the electricity supply authority, a stand-alone system is not viable for areas where grid power is available. Substantial reductions in the cost of electrical energy storage on a small scale would be required. This may be possible with systems such as the redox system described in Chapter 12.

Without an inexpensive method of storing electricity, the most viable approach is to connect the photovoltaic system to the electricity supply grid. This eliminates the need for *long-term* energy storage. Several different system configurations are possible in this grid-connected mode. An integral part of each of these systems is an inverter to convert solar cell dc output to alternating form.

Even though the grid may be used as a long-term storage medium, the question as to whether *short-term* storage on site is desirable has to be addressed. This storage would help tide over the system overnight or through short periods of bad weather. For prolonged periods of poor weather, the presence of short-term storage would allow the grid to supply power to the residence at its convenience. A system without on-site storage is also workable, particularly if the utility is prepared to buy back excess electricity generated. The optimum size of the array in this case increases for increasing buy-back rates.

From the householder's point of view, the optimum system depends on the relative costs of solar and storage systems, plus the utility rate structure. Large differentials in rates depending on the time of day would encourage increased storage, whereas a willingness on the utility's part to buy back excess electricity generated by the residence at a reasonable fraction of selling rates would decrease the optimum storage size. Battery storage is the most promising storage technique. The major disadvantages are the potential hazards associated with batteries in residential settings and problems posed by the regular maintenance the batteries would require. Nonetheless, with suitable venting and electronic protection of the battery, such storage is feasible (Ref. 14.1). A possible battery-pack configuration is shown

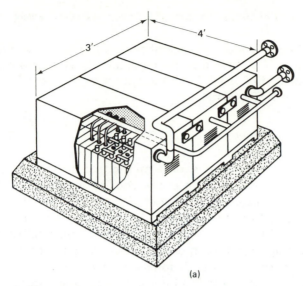

200-V system
20 kWh
80% energy efficiency
5000 cycles

Voltage
  Max. charge, 250
  Min. discharge, 180
Cells
  Number, 96
  Height, 1.5 ft.
Battery
  Area, 12 ft^2
  Weight, 3000 lb

(a)

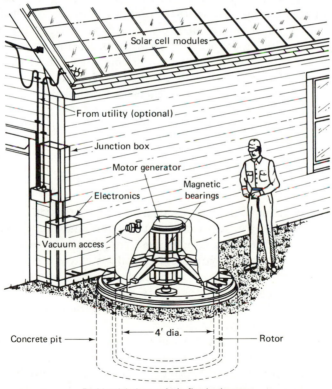

Residential photovoltaic flywheel system

(b)

Figure 14.1. Energy storage concepts for residential systems:

(a) Possible battery configuration. (After Ref. 14 1.)
(b) Flywheel storage. (After Ref. 14.2, © 1980 IEEE.)

in Fig. 14.1(a). A redox system as described in Chapter 12 would have several advantages if successfully developed. Flywheel storage has also been considered (Ref. 14.2). An indication of the size of the unit required is given in Fig. 14.1(b).

14.2.2 Module Mounting

Studies indicate that the cheapest method of mounting the solar modules is to integrate them into the roofing as indicated in Fig. 14.2(a) so that they form the dual role of generating electricity and providing protection from the elements (Ref. 14.3). Retrofitted modules probably would be mounted in the stand-off configuration of Fig. 14.2(b). Although installation costs would be similar for both configurations, there would be an additional credit for normal roofing materials displaced by the solar modules in the integral approach.

Optimum module size for this application has been estimated to be about 0.8 × 2.5 m with a corresponding module weight of about 25 kg (Ref. 14.3). Wiring costs decrease with increasing array voltage output, although this dependence is not strong for voltages over 100 V dc (Ref. 14.3). The reference cited indicates that aesthetic considerations favor rectangular modules with aspect ratios about 2:1, of a dark earth-tone color, and having a matt finish. Shingle modules [Fig. 14.2(c)] may also be attractive if inexpensive interconnection techniques can be developed (Ref. 14.4).

14.2.3 Thermal Generation

A large portion of the residential use of energy will be in the form of low-grade heat for hot-water systems and space heating. The question arises as to how this can be best supplied in a system that includes photovoltaic cells. Three possible alternatives are to use the photovoltaic modules to provide both electrical and thermal requirements of the residence, to use a separate solar thermal collector for supplying the thermal loads, or to use combined photovoltaic/thermal modules or a *total energy system.*

Although conceptually appealing, the total energy system with photovoltaic and thermal functions incorporated into the same module has a number of disadvantages. The solar cells of necessity will operate at higher temperatures and hence lower efficiency. The thermal collector will also operate at lower efficiency, because the cells will be extracting some of the available energy. The ratio of ther-

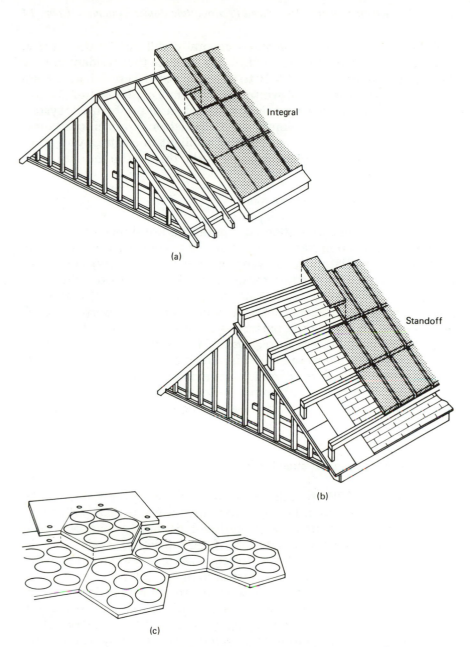

Integral

Standoff

(a)

(b)

(c)

Figure 14.2.   Possible mounting schemes for rooftop solar
cell modules:

(a)   Integral. (After Ref. 14.3.)
(b)   Standoff. (After Ref. 14.3.)
(c)   Shingle.

253

mal to electrical energy generated generally will not be the same as that required by the residence. Studies indicate that residential total energy systems seldom are likely to be more cost effective than systems with optimum areas of photovoltaic and thermal modules (Ref. 14.5).

The choice between this system and an all-photovoltaic system depends on whether the simplicity of the latter approach overcomes the inefficiency inherent in converting sunlight to electricity and then to heat. At low photovoltaic module costs, this can be the case.

14.2.4   System Configurations

Several possible system configurations are shown in Fig. 14.3. In the first, shown in Fig. 14.3(a), battery storage is used in conjunction with a regulator (to prevent battery overcharging) connected between the module and the battery. In this connection, the input voltage to the inverter is the battery voltage. Fig. 14.3(b) shows a potentially more efficient variation where only that portion of the output used to charge the battery is diverted through the regulator branch. Without battery storage as in Fig. 14.3(c), the output of the module is connected to the inverter, which can be designed to ensure that the module delivers power at its maximum power point. Finally, Fig. 14.3(d) shows how a thermal collector could be integrated into the previous system. In each case, the ac output of the inverter is synchronized to the utility input.

14.2.5   Demonstration Program

In late 1979, the U.S. Department of Energy initiated a Solar Photovoltaic Residential Project. The purpose was to clarify the issues involved in the residential use of photovoltaics in grid-connected areas and to facilitate commercialization of appropriate systems. As originally outlined (Ref. 14.6), the project was to be completed in 1988, involving three major stages as illustrated in Fig. 14.4.

The first stage involved the construction of regional experiment stations in representative areas of the country. These stations were to test prototype systems designed and built by industry. These prototype systems were roof-only systems installed at the stations, delivering the estimated electrical and thermal residential requirements. The actual use of these quantities by residences in the neighborhood of the station was monitored at the same time as the performance of the prototypes to check compatibility.

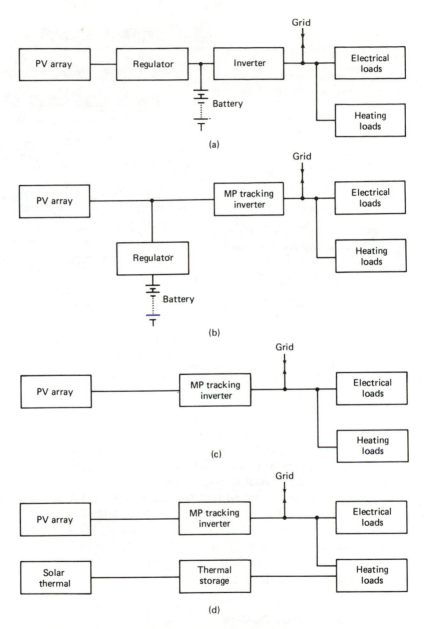

Figure 14.3. Possible interconnection scheme for photo-voltaic residential systems. See the text for details..

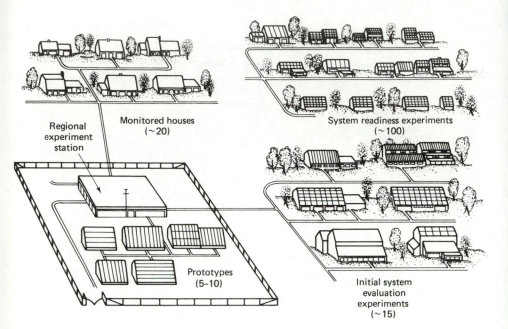

Regional experiment station

Monitored houses (~20)

System readiness experiments (~100)

Prototypes (5-10)

Initial system evaluation experiments (~15)

Figure 14.4. Outline of the DOE Residential Program, ranging from initial prototype testing in regional experiment stations to the deployment of a large number of solar-cell-powered residences in the commercial readiness experiments. (After Ref. 14.6.)

In the next stage, refinements of successful prototypes were to be tested by installation at a small number of occupied residences near the experiment stations. These initial system-evaluation experiments were concerned not only with physical performance but also with occupant and institutional responses.

In the final phase, scheduled to start in 1984, system readiness experiments were to be conducted independently of the regional experiment stations. In this phase, clusters of about 100 solar-powered residences were to be built to confront, head-on, the institutional and engineering issues resulting from the widespread use of photovoltaics in the residential sector (Ref. 14.7).

14.3 CENTRAL POWER PLANTS

14.3.1 General Considerations

The ultimate goal for photovoltaics is to compete in an economic sense with conventional methods of generating large quantities

256

of electricity in a centralized power plant. Several studies have identified essential properties for this to occur.

One obvious condition is that the solar modules must be inexpensive, somewhat less expensive than for residential use. A less
obvious condition is that the modules have to be efficient. A solar
array efficiency of at least 10% is desirable. Lower efficiencies increase
the area of the array that is required to produce a given output. This
increases such costs as site preparation, the cost of support structures,
installation costs, and maintenance costs. These costs, as well as those
of power conditioning equipment, are often referred to as *balance-of-
system* costs. Careful consideration has to be given to keeping balance-
of-system costs to a minimum.

Studies indicate that the optimum module size for large-scale
application is of the order of 1.2 × 2.4 m (Ref. 14.8). A large number
of different approaches have been considered for support structures.
The most severe load placed on this support structure is wind loading,
with the design wind loading largely determining its cost. The low
profile of photovoltaic arrays and the aerodynamic shielding provided
by adjacent rows in an array field or by a surrounding fence act to
greatly reduce these loads. Preliminary results show that the wooden-
pole support system of Fig. 14.5(a) looks attractive for smaller installations, and that the concrete-truss support system of Fig. 14.5(b)
becomes competitive for large installations (Ref. 14.9).

It is unlikely that a terrestrial photovoltaic system would be
used as the *sole* source of electricity in any grid network. Either the
costs of long-term energy storage would be prohibitive, or the array
would have to be grossly oversized so that it could generate the required energy even in cloudy weather. It would be possible to alleviate
this difficulty by using the photovoltaic system in conjunction with a
nonsolar plant of lower power rating together with relatively short
term storage. The most likely role that photovoltaics would play in a
large network in the foreseeable future is a fuel-displacement role.
Since such a network already has considerable flexibility in meeting a
time-varying demand, photovoltaics can be used *without storage*, provided that they contribute less than about 10% of the total generating
capacity of the network.

Because of the diffuse nature of solar radiation, large land areas
are required to generate substantial amounts of energy using photovoltaics. Some perspective to the question of land use is given by
considering the percentage of land required to be covered by photovoltaics to generate *all* the energy requirements of selected countries.
The results of such a calculation are shown in Table 14.1. Although
some European countries have poor results, for obvious reasons, in

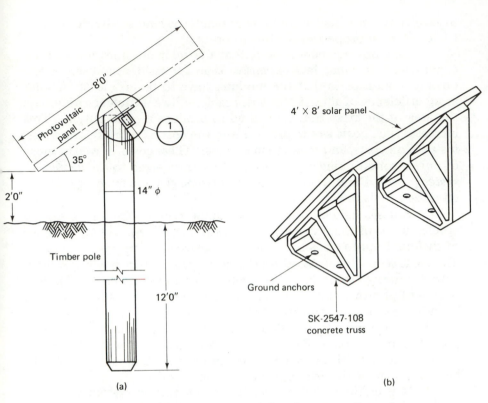

Figure 14.5. Alternative approaches for supporting solar cell modules in large field installations:
(a) Wood pole/torsion tube system.
(b) Concrete truss system.
(After Ref. 14.9, © 1980 IEEE.)

many countries such as the United States the required land area is less than that presently covered by man-made structures such as buildings and roads. Although an enormous task, the installation of a photovoltaic system over several decades capable of supplying all the world's energy requirements obviously is not beyond present engineering realities.

14.3.2 Operating Mode

Figure 14.6(a) shows a plot of the power demand placed on a hypothetical electricity authority during a typical day. Also shown is the effect that a solar plant incorporated into the authority's network

Table 14.1. PERCENTAGE OF LAND AREA REQUIRED TO GENERATE
ALL THE ENERGY REQUIREMENTS OF THE COUNTRIES SHOWN IN
1970 WITH PHOTOVOLTAICS OF 10% EFFICIENCY

Australia	0.03
Canada	0.20
Denmark	4.5
Eire	1
France	3.5
Israel	2.5
Italy	4
Netherlands	15
Norway	0.50
South Africa	0.25
Spain	1
Sweden	0.75
United Kingdom	8
United States	1.5
West Germany	8

Source: After D. O. Hall, "Will Photosynthesis Solve the Energy Problem?" in *Solar Power and Fuels*, ed. J. R. Bolton (New York: Academic Press, 1977), p. 36.

would have upon the load to be supplied by the remainder of the system. In a system where the load demand peaks in the evening, the effect of the solar plant is to sharpen the peak and broaden the trough of the daily demand curve.

As previously mentioned, with small penetration of the supply network, no storage is necessary with photovoltaics. The grid will be capable of absorbing the fluctuating solar supply just as it now absorbs fluctuating demand. Without storage, the solar plant would act as a fuel saver, decreasing the time that *intermediate* and *peaking* generating equipment would have to operate. However, the tendency to sharpen peaks in the daily demand curve improves the effectiveness of a given amount of energy storage in leveling demand, as shown schematically in Fig. 14.6(b) and (c). Thus, even though the energy storage is charged primarily by the *base-load* equipment in the net work rather than the solar plant as indicated in Fig. 14.6(b), there is a synergistic effect between solar plant and energy storage. The presence of one improves the performance and viability of the other. It follows that networks installing energy storage now for operation with coal- and nuclear-base-load plants will be in the best position to implement solar plants as they become available.

Another important concept in central power plant operation is the *capacity credit* to be given to solar plants. It might be argued

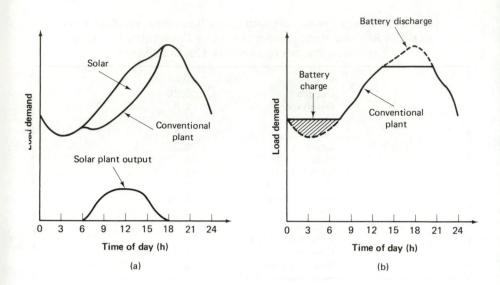

(a)

(b)

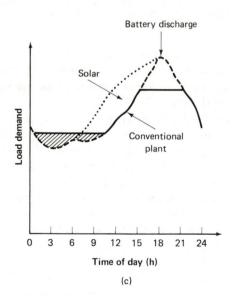

(c)

Figure 14.6. (a) Typical daily load demand profile, showing the effect of a small contribution from a solar plant. (b) Use of energy storage to trim the peak in the load demand profile for a system without solar input. (c) As for part (b) except with solar input.
(After Ref. 14.10, © 1978 IEEE.)

that no credit could be given to a photovoltaic plant because its output would be very low on cloudy days and backup capacity would be required to cover these days. The actual situation is more complex. Conventional equipment can also be forced out of service unexpectedly. A method used to calculate the capacity of a system is to specify the level of reliability in meeting demand that is required and to calculate the maximum load that can be met to this level of reliability.

Figure 14.7 shows the result for a particular grid network of such a calculation for a base system without photovoltaics, for a system that includes 500 MW$_p$ of photovoltaics, and for a system with the latter as well as 2000 MWh of battery storage (Ref. 14.10). In this system the capacity credit for photovoltaics works out at about one-third of peak capacity. Combined with the battery storage, the capacity credit increases to 580 MW. In this example, the maximum power demand occurs at about 6 P.M. in summer as is typical of many U.S. utilities. If the peak occurs closer to midday, the capacity credit for photovoltaics alone would be even a higher fraction of peak capacity. On the other hand, for systems where the peak load occurs outside the daylight hours (e.g., winter evenings), the capacity credit would be much smaller.

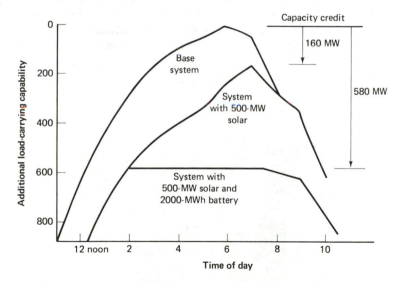

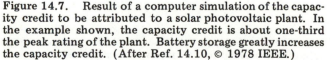

Figure 14.7. Result of a computer simulation of the capacity credit to be attributed to a solar photovoltaic plant. In the example shown, the capacity credit is about one-third the peak rating of the plant. Battery storage greatly increases the capacity credit. (After Ref. 14.10, © 1978 IEEE.)

14.3.3 Satellite Solar Power Stations

No book on photovoltaics would be complete without some mention of the imaginative concept of using large solar cell arrays in space to capture sunlight and beam energy back to earth as microwaves, as illustrated in Fig. 14.8. The solar station would be placed in geosynchronous orbit around the earth at an altitude that is large compared to the earth's radius. This would ensure that the earth did not shadow the array except for periods of about 1 h around local midnight for the weeks surrounding the equinoxes. A more detailed discussion is given in Ref. 14.11.

The major advantage of the system is that sunlight would be available continuously except for the periods mentioned above. Storage is not required and the system could be used in a base-load role. Other advantages include the higher sunlight intensities in space and the fact it is relatively easy to ensure the arrays are always nearly normal to the sun's rays. A solar array of a given peak rating could generate five to eight times the energy of a similarly sized array at a sunny location on earth. This factor would be reduced by the losses associated with the transmission of the collected energy to earth.

The major disadvantages are strategic vulnerability and high balance-of-system costs. As well as the immense problems associated with constructing such an array in space and maintaining it, a collecting receiver of significant size has to be constructed on earth. The possible environmental effects of a microwave beam of the required size and intensity also deserve careful scrutiny.

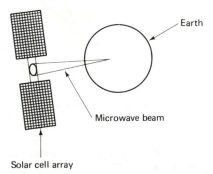

Figure 14.8. Satellite solar power station. The energy collected is beamed back to earth as microwaves.

In this, the final chapter of this book, issues surrounding two proposed long-term applications of solar cells have been considered. The costs required for cells to be viable in these applications are, without a doubt, within the realms of technical feasibility.

Residential use of photovoltaics in grid-connected areas becomes possible at higher module costs than central power generation using photovoltaics because the appropriate point of view to calculate costs in the former case is that of the householder. From the technical point of view, several solar cell technologies have been recognized which can produce cells at the costs required for this application. The major obstacles to the residential use of photovoltaics are not technical but institutional. These issues relate to building codes, integration with utilities, financing of the solar arrays, and so on.

Photovoltaics are unlikely to be used in a stand-alone mode in a central power-generating role because of the excessive energy storage required. With small but significant contributions to the grid-generated power, energy storage is not an essential partner to a photovoltaic plant. The two elements do complement each other to some extent, the presence of either making the other more viable. In conjunction with a nonsolar backup system and moderate amounts of storage, ultimately photovoltaic plants would be capable of generating the majority of power supplied by a grid network.

Thin-film technology is available at the present point in time for producing solar cells at the costs *per unit area* required for this central power-generating role. The improvement required is in the power output per unit area obtained from such cells. Expressed in a different way, thin-film amorphous silicon cells became commercially available in 1980 whose performance equaled that of high-quality single-crystal cells for the indoor applications for which they were designed. The challenge facing photovoltaics is to produce thin-film cells that can duplicate this feat *outdoors*.

REFERENCES

[14.1] W. FEDUSKA et al., *Energy Storage for Photovoltaic Conversion; Residential Systems—Final Report*, Vol. 3, prepared for US National Science Foundation, Contract No. NSF C-7522180, September 1977.

[14.2] A. R. MILLNER AND T. DINWOODIE, "System Design, Test Results, and Economic Analysis of a Flywheel Energy Storage and Conversion System for Photovoltaic Applications," *Conference Record, 14th IEEE Photovoltaic Specialists Conference, San Diego*, 1980, pp. 1018–1024.

[14.3] P. R. RITTELMANN, "Residential Photovoltaic Module and Array Requirements Study," in *Proceedings of the U.S. DOE Semi-Annual Program Review of Photovoltaics Technology Development, Applications and Commercialization*, U.S. Department of Energy, Report No. CONF-791159 (1979), pp. 201–223.

[14.4] N. F. SHEPARD, JR. AND L. E. SANCHEZ, "Development of a Shingle-Type Solar Cell Module," *Conference Record, 13th IEEE Photovoltaic Specialists Conference, Washington, D.C.*, 1978, pp. 160–164.

[14.5] V. CHOBOTOV AND B. SIEGAL, "Analysis of Photovoltaic Total Energy System Concepts for Single-Family Residential Applications," *Conference Record, 13th IEEE Photovoltaic Specialists Conference, Washington, D.C.*, 1978, pp. 1179–1184.

[14.6] M. D. POPE, "Residential Systems Activities," in *Proceedings of the U.S. DOE Semi-Annual Program Review of Photovoltaics Technology Development, Applications and Commercialization*, U.S. Department of Energy, Report No. CONF-791159 (1979), pp. 346–352.

[14.7] E. C. KERN, JR. "Residential Experiments," in *Proceedings of Photovoltaics Advanced R and D Annual Review Meeting*, Solar Energy Research Institute Report No. SERI/TP-311-428 (1979), pp. 17–36.

[14.8] P. TSOU AND W. STOLTE, "Effects of Design of Flat-Plate Solar Photovoltaic Arrays for Terrestrial Central Station Power Applications," *Conference Record, 13th IEEE Photovoltaic Specialists Conference, Washington, D.C.*, 1978, pp. 1196–1201.

[14.9] H. N. POST, "Lost Cost Structures for Photovoltaic Arrays," *Conference Record, 14th IEEE Photovoltaic Specialists Conference, San Diego*, 1980, pp. 1133–1138.

[14.10] C. R. CHOWANIEC et al., "Energy Storage Operation of Combined Photovoltaic/Battery Plants in Utility Networks," *Conference Record, 13th IEEE Photovoltaic Specialists Conference, Washington, D.C.*, 1978, pp. 1185–1189.

[14.11] D. L. PULFREY, *Photovoltaic Power Generation* (New York: Van Nostrand Reinhold, 1978), pp. 56–62.

Appendix

PHYSICAL CONSTANTS

q electronic charge = 1.602×10^{-19} coulomb

m_0 electronic rest mass = 9.108×10^{-28} g
$$= 9.108 \times 10^{-31} \text{ kg}$$

c velocity of light in vacuum = 2.998×10^{10} cm/s
$$= 2.998 \times 10^{8} \text{ m/s}$$

ϵ_0 permittivity of free space = 8.854×10^{-14} farad/cm
$$= 8.854 \times 10^{-12} \text{ farad/m}$$

h Planck's constant = 6.625×10^{-27} erg-s
$$= 6.625 \times 10^{-34} \text{ joule-s}$$

k Boltzmann's constant = 1.380×10^{-16} erg/K
$$= 1.380 \times 10^{-23} \text{ joule/K}$$

$\dfrac{kT}{q}$ thermal voltage = 0.02586 V (at 300 K)

λ_0 wavelength in vacuum associated with photon of 1-eV energy
$$= 1.239 \text{ } \mu\text{m}$$

PREFIXES

milli (m) = 10^{-3} kilo (k) = 10^{3}
micro (μ) = 10^{-6} mega (M) = 10^{6}
nano (n) = 10^{-9} giga (G) = 10^{9}
pico (p) = 10^{-12}

Appendix

SELECTED PROPERTIES OF SILICON (AT 300 K)

E_g energy gap = 1.1 eV (see Table 3.1)

N_C effective density of states in conduction band
= 3×10^{19} cm^{-3} = 3×10^{25} m^{-3}

N_V effective density of states in valence band
= 1×10^{19} cm^{-3} = 1×10^{25} m^{-3}

n_i intrinsic concentration = 1.5×10^{10} cm^{-3} = 1.5×10^{16} m^{-3}

ϵ_r relative permittivity = 11.7

$\hat{n}$ refractive index = 3.5 (at a 1.1-μm wavelength) (see Fig. 3.1)

μ_e electron mobility $\leq$ 1350 cm^2/V-s $\leq$ 0.135 m^2/V-s [see Eq. (2.36)]

μ_h hole mobility $\leq$ 480 cm^2/V-s $\leq$ 0.048 m^2/V-s [see Eq. (2.36)]

D_e electron diffusion coefficient = $0.02586\mu_e$

D_h hole diffusion coefficient = $0.02586\mu_h$

ρ electrical resistivity [see Eq. (2.35)]
density = 2.33 g/cm^3 = 2330 kg/m^3

LIST OF SYMBOLS

ξ	electric field strength
α	absorption coefficient
ϵ	dielectric constant
ϕ	work function
ϕ_B	barrier height
λ	wavelength
μ	mobility
η	efficiency
ρ	space-charge density; resistivity; sheet resistivity; specific contact resistance
σ	conductivity
τ	lifetime
ψ_0	built-in potential
χ	electron affinity
A	cross-sectional area
c	velocity of light in vacuum

D	diffusion coefficient
E	energy
E_c, E_v	energies of conduction- and valance-band edges
f_c	collection probability of light-generated carriers
E_F	Fermi level
FF	solar cell fill factor
G	generation rate of electron–hole pairs
h	Planck's constant
I	current; intensity
I_0	diode saturation current
I_{sc}	short-circuit current
J	current density
J_e, J_h	electron and hole current densities
k	Boltzmann's constant
$\hat{k}$	extinction coefficient
L_e, L_h	diffusion length for electrons and holes
m_0	electronic rest mass
m_e^*, m_h^*	effective mass of electrons and holes
n	electron concentration
n_{n0}, n_{p0}	thermal equilibrium concentration of electrons in n-type and p-type semiconductors
$\hat{n}_c$	refractive index
$\hat{n}$	real part of refractive index
n_i	intrinsic concentration
N_C, N_V	effective densities of states in conduction and valance bands
N_A, N_D	densities of acceptors and donors
p	hole concentration; crystal momentum; fractional power loss
p_{n0}, p_{p0}	thermal equilibrium concentration of holes in n-type and p-type semiconductors
q	electronic charge
R	resistance
t	time
T	temperature
U	net recombination rate
V	voltage; potential
V_{oc}	open-circuit voltage

BIBLIOGRAPHY

BACKUS, C. E., ed., *Solar Cells.* New York: IEEE Press, 1976. A collection of technical papers significant in the development of solar cells.

HOVEL, H. J., *Solar Cells*, Vol. 11, Semiconductor and Semimetal Series, ed. R. W. Richardson and A. C. Beer. New York: Academic Press, 1975. A review of the theory and performance of solar cells.

JOHNSTON, W. D., *Solar Voltaic Cells.* New York: Marcel Dekker, 1980. Review of the current status of photovoltaic development.

MERRIGAN, J. A., *Sunlight to Electricity: Prospects for Solar Energy Conversion by Photovoltaics.* Cambridge, Mass.: MIT Press, 1975. Investigates the technical practicality and economic viability of solar cells.

NEVILLE, R. C., *Solar Energy Conversion: The Solar Cell.* Amsterdam: Elsevier, 1978. Emphasis on the theoretical effects of relevant parameters on solar cell performance.

PULFREY, D. L., *Photovoltaic Power Generation.* New York: Van Nostrand Reinhold, 1978. Treatment of the technical, economic, and institutional issues relevant to the large-scale terrestrial application of solar cells.

RAUSCHENBACH, H. S., *Solar Cell Array Design Handbook.* New York: Van Nostrand Reinhold, 1980. Source of practical data related to solar cell module and array design for terrestrial and space systems.

SITTIG, M., *Solar Cells for Photovoltaic Generation of Electricity.* Park Ridge, N.J.: Noyes Data Corporation, 1979. A guide to U.S. patent literature in the photovoltaic field between 1970 and 1979.

INDEX